Human Anatomy Lab Manual

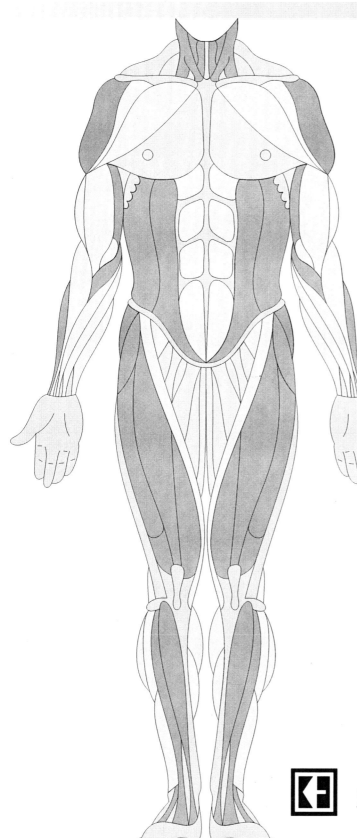

Corey S. Johnson

*University of North Carolina–
Chapel Hill*

KENDALL/HUNT PUBLISHING COMPANY

4050 Westmark Drive Dubuque, Iowa 52002

Contents

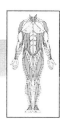

Chapter 6   225
Central Nervous System: The Brain

Chapter 7   229
The Heart and Its Circulation

Chapter 8   239
The Respiratory System

Chapter 9   245
Digestive Organs

Chapter 10   253
Urinary and Reproductive Organs

Introduction

How to Use This Manual

The study of anatomy can simply be the memorization of the names of structures in the body. Surely, this is important, for we must all learn the language of a new subject. However, just like learning anything else, we must progress from the simple to the complex. Your study of anatomy will at first focus on learning anatomical terminology, progress to a focus on naming structures, and finally arrive at a comprehensive view of the body in three dimensions that includes an understanding of structure/function relationships.

This manual is designed to present information visually, because most people learn this way and because anatomy is a visual science. The illustrations are simple line drawings rather than colorful illustrations or photographs with lines and labels exploding every which way. Such illustrations are intended to convey as much information as possible. The approach I have taken here is to present many illustrations that each convey very little information so that the student can build up their understanding from smaller, bite-sized pieces. These illustrations, largely by Jamey Garbett, will undoubtedly demonstrate the simplest view of a structure within the larger context of the body or skeleton.

It is highly recommended that you keep a set of colored pencils with your manual and lightly color-code various structures. Like taking notes in lecture, the simple act of coloring the structure (with its name and function in mind) will help solidify the term in your mind. Also, if you ever doubt that you learned a structure, take a look and see if it's colored in! More complex illustrations are available from a variety of sources, including your textbook and those in your lab. Use this resource as a means to study structures within the larger context of the human body.

What Can I Expect from Lab?

Labs will consist of: (1) a short pre-lab quiz; (2) a period of instruction—lecture, demonstration, or otherwise; (3) examination of models, bones, preserved organs, or preserved cat specimens; and (4) a period of time for review.

Anatomy is taught either by system (skeletal, muscular, nervous systems) or by region (arm, leg, abdomen). In this laboratory, you will be examining anatomy by system, but those systems will be grouped into regions. For example, in one laboratory you will study the bones of the body axis: the skull, vertebral column, and the ribs. Then, you will examine the muscles that attach to these bones. In a later lab you will study the contents of the thoracic cavity (cardio-vascular and respiratory systems). Then, you will study the contents of the abdominopelvic cavity (the digestive, urinary, and reproductive systems).

Why use models and animal organs? The organs used in this laboratory closely approximate the human counterparts. Additionally, physically manipulating an object gives one a sense of the object more than just looking at a picture of it. People don't usually buy pictures of sports cars: they drive them! Anatomy is three-dimensional, so we will spend most of our time looking at three-dimensional things. Sometimes models are useful where animal models are less than ideal. The advantage, even over human cadaver dissections, is that models can show structures clearly, where a comparable cadaver dissection showing the same structures would take untold hours of preparation. Computer-based anatomy instruction is no longer used in this course, as many students and instructors find it fails to convey three-dimensional relationships. However, for those who have come to the point of comprehending the spatial relationships in the body, a number of excellent programs are available for purchase online including ADAM®, Real Anatomy®, and Anatomy and Physiology Revealed®.

What Will Exams Be Like?

Pre-lab quizzes: To ensure a proper background for the day's lab, pre-lab quizzes will be given. Your TA will explain the format of these exams on the first day of lab. They will likely consist of 10 questions that test your recognition of specific structures.

Laboratory exams: Laboratory exams will be given to test your knowledge about anatomy. You can expect two sections on each exam. First is the identification of structures. You will be presented with a structure and be expected to name the structure (with reasonable spelling). A "word list" is provided for each laboratory and from these lists, 25 terms will be selected for identification. An additional 25 questions will be asked that require a short answer. These questions will test your understanding of function and three-dimensional relationships.

How Can I Do Well in This Lab?

This lab course and preparations for it should take about ¼ of the time spent in lecture and preparing for its exams. Some will find lab very easy, and it would benefit you greatly to identify these people and imitate their study habits. Others of us will find great difficulty in recalling terms and visualizing structures. To start with, you must define the limits of what you must know. Most of this has been done for you. For each term in the word list, (1) visualize the structure, (2) understand why it is named the way it is (your TA should be able to help you with this), (3) learn its function or spatial relationships, and (4) return to your visualization and incorporate these other aspects.

Another tip that is incredibly useful for this and other classes: study the way in which you will be tested. Once you have learned the anatomy, it's time to remember it. Your recall will only be as good as your study habits. If your study does not in some way replicate the testing situation, you will do poorly. Since the tests will require you to pull a term from your knowledge based on what you see, you should spend some time preparing for the test in this way. Find a source of images (http://daphne.palomar.edu/ccarpenter/Models/model%20index.htm, for example) and quiz yourself. Also, there are software packages that do this as well. I like *Practice Anatomy Lab* by Benjamin Cummings. See the structure from several vantage points such as illustrations, photographs, and models so that your visualization (remember steps 1 and 4 above) is accurate.

Sometimes there is just plain old-fashioned rote memorization. For times like these, use a mnemonic device to help you remember it. Medicalmnemonics.com has a searchable database if you need some ideas. As an example, the wrist (carpus) has eight bones. To remember the order, you might instead remember the phrase: "**S**imply **L**earn **T**he **P**arts **T**hat **T**he **C**arpus **H**as." And you may remember the proper order: **S**caphoid, **L**unate, **T**riquetrium, **P**isiform, **T**rapezium, **T**rapezoid, **C**apate, **H**amate. Sometimes the mnemonic is harder to remember than the anatomy, but remember: nobody cares *how* you learn it, so long as you *do* learn it.

Anatomical Nomenclature

History

"I have seen further because I stood on the shoulders of giants."
—Sir Isaac Newton

Do you know that I know more anatomy than any of my more famous anatomical predecessors? In fifteen weeks you also will know more anatomy than many of the great early anatomists. This is only true, however, because of their efforts. Newton's famous quote reminds us of these efforts, saying much about the history of scientific knowledge. We should never lose sight of this fact. Today we live in a world where the knowledge available to us is unfathomable and is accumulating at an unparalleled rate. Yet, it is important to remember that it is cumulative. Many individuals have contributed to the knowledge base which we now possess. We only know all that we know because of the great minds that laid the foundations and stimulated further thought. Many have contributed and many will continue to contribute. As the great anatomist Vesalius said, "Vivitur ingenio, caetera mortis erunt," or genius lives on, all else falls dead in silence.

The Anatomical Instinct

When did our interest in anatomy begin? Who were the first anatomy scholars? The anatomical historian Charles Singer, in his book *The Evolution of Anatomy*, suggests that there is an anatomical instinct. He writes, "To reach its first beginnings we should have to carry our search far back indeed. Nor could we safely stop even with humankind. We might discern rudiments of it in the neck-breaking clutch of the tiger or in the accurate puncturing by the ichneumon-fly of the ventral ganglia of its victim." Even in our own species we see evidence of an early knowledge of anatomy, as prehistoric hunters recorded on cave walls illustrations of arrows piercing the hearts of their prey. While this intrinsic or utilitarian knowledge of anatomy seems ancient, its realm as a science is a recent phenomenon in the great span of human history. For only the relatively short span of a few thousand years have anatomical conceptions been consciously formulated and consciously accumulated. Yet during this relatively brief time, a rich and interesting history has emerged. A story with a fascinating cast characters, ranging from the great Renaissance artists to numerous early scientists who robbed graves in search of anatomical knowledge. The result from these years of inquiry is a massive knowledge base, which we use, study, refine, and continue to supplement with our own research and knowledge.

Anatomy and Us

The great anatomical knowledge base that has accumulated over the years provides us with much information about our body's structure and function. Many professions base their existence on this scholarship. While there are many individuals well versed in this knowledge, it is sad but true that there exist many more with little understanding of this most prized possession—their body. Is it not enigmatic that as we pass through the halls of education we are required to learn many things, yet there is no requirement to learn about the one thing we will never be without—our body? For example, we must all take and pass English classes, writing classes, and history classes in order to leave the portals of higher education with a degree. While I do not disagree with these requirements, it is sad that there is no requirement for us to learn about our own body, a constant companion for life. A little insight to the structure and workings of this wonderful machine can pay great dividends. It can help us become more aware of what we can do to better insure its longevity. It can help us communicate more clearly with health professionals about our health concerns. More importantly, it can facilitate our understanding of the concerns they share with us. It can help us avoid those products and people interested in making a fast buck off our ignorance. So I congratulate you for your interest in your body. I guarantee you that you will not regret the knowledge you attain.

The Anatomical Legacy

Because anatomy is an ancient science its roots reside in the languages of the early anatomy scholars, the languages of Latin and Greek. As a result of this Latin and Greek legacy, the modern student of anatomy faces a daunting task, that of learning anatomical terminology—a foreign language. And learn it we must if we are to understand anatomy. Today as we embark upon the study of anatomy, we unknowingly accumulate a large Latin and Greek vocabulary. This plethora of terminology is often cursed by beginning anatomy students, as they are forced to learn this "foreign tongue." However, the language need not be a curse, but rather a blessing in disguise. If you will learn to analyze unfamiliar Greek and Latin anatomical terms in light of their English equivalent, the logic of the terminology will usually become apparent. Most of the Greek and Latin terms are very descriptive adjectives that the English language has converted into nouns. Using this approach, you can associate unfamiliar words with familiar English terms that conjure up great visual images. This association process facilitates memorization. Many studies have demonstrated that learning by association enhances long-term memory. This is important, as the old maxim says, "it is not what you know, but what you remember that makes you wise." The next section demonstrates this process. Because this can be such an effective technique, I want to minimize the amount of effort it requires of you to utilize it. With this in mind, every time a new term occurs in the book, the term will be accompanied by its etymology. Also, included at the back of this manual is an alphabetical list of anatomical terms with their Latin and Greek derivations. This list will provide you quick access to the etymology, which can help you learn and memorize through the use of this powerful association technique.

Samples of the Use of Etymology
Acromion

Today, if we were to logically describe the landmark known as the acromion, we might call it the bone at the tip or end of the shoulder. This is exactly what we do when we say the word acromion. The English word acromion is a combination of the original Greek words akros meaning tip or peak and omos which translates as shoulder. Therefore, we make an important association. We realize that when we say acromion, we are simply saying the tip of the shoulder. Additionally, it is comforting to realize that the anatomists of old named things in a logical manner. It is a pleasing scenario to envision a few ancient anatomists sitting around a table discussing names for the landmarks of the scapula. They recognize this landmark as the projection of bone that they can touch at the edge of their shoulder. One says to the others, "Let's call it the bone at the tip or peak of the shoulder," or in Greek "os akros omos." The others respond, "Great name!" And so it is. The only problem—they used the Greek terms, because they did not speak English. As the Greek anatomical words were assimilated into the English language, the Greek noun "os" meaning bone was dropped and the adjective, "akros" meaning peak and the genitive, "omos" meaning shoulder, were combined to form the new English noun "acromion."

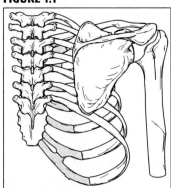

FIGURE 1.1

Illustration by Jamey Garbett. © 2003 Mark Nielsen.

Pudendal Nerve

The etymology behind some words is actually somewhat amusing, this is the case with the word pudendal. This is an anatomical term that refers to the external genitalia. For example, there is an internal pudendal artery and vein, as well as a pudendal nerve. These structures supply blood to and innervate the genital regions of the body. In referring to the genitalia, the old anatomists used the Latin term pudenda, which translates into the English as "that which you should be ashamed of." Apparently these anatomists were fairly modest sorts, not the type of people we would see frequenting nudist colonies.

FIGURE 1.2

Illustration by Jamey Garbett.
© 2003 Mark Nielsen.

Latissimus Dorsi

It should be understood that all the muscles of the body have the word muscle in their name, such as the latissimus dorsi muscle, biceps brachii muscle, and rectus abdominis muscle. Often, however, we eliminate the word muscle in the English nomenclature and latissimus dorsi, for example, stands alone as the noun in the English language. However, in the original Greek and Latin nomenclature the word "musculus" is the noun, and the words "latissimus" and "dorsi" represent the adjective and genitive, respectively, that describe the noun "musculus." As with all adjectives and genitives, these words are descriptive modifiers of the noun that help us form visual images of the noun, in this case of "musculus" or the muscle. In this instance, latissimus translates as "the broadest" and dorsi as "the back." The result is, "the broadest muscle of the back," a very apt description of the muscle.

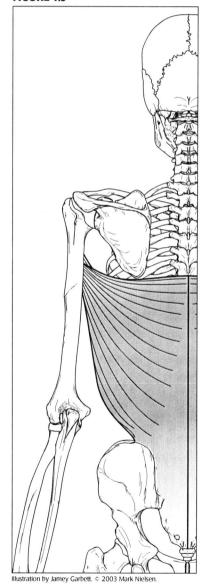

FIGURE 1.3

Anatomical Nomenclature

As a descriptive science, anatomy is a subject which places a large emphasis on description. As a result, a large database of names have accumulated during the last few thousand years of anatomical history. Ideally one wants to be able to use terms that are precise, simple, and internationally recognized. With this purpose in mind, anatomists, beginning in the year 1895, established a standardized Latin vocabulary of anatomical terminology. Continuous revisions of this terminology over the past century have led to the current dictionary of terms, defined by the Federative Committee on Anatomical Terminology (FCAT) as the *Terminologia Anatomica*. (Published in 1998 by Thieme Publishers.) This book lists both the accepted Latin term and its English equivalent. In this lecture manual and throughout the anatomy course, I will adhere to the terms established by the *Terminologia Anatomica*. Where the *Terminologia Anatomica* recognizes synonyms I will list both terms when first listed, for example, the English subcutaneous tissue has the Latin equivalents of either tela subcutanea or hypodermis. If you encounter other terms in books, such as superficial fascia or subcutaneum, you should ignore them and learn to use the proper terms.

Terms of Position/Comparison/Relationship

To help improve communication when discussing the basic parts of the body and the relationships those parts have to one another, there are some important terms with which you should become familiar. These are terms that describe the position, relation, and orientation of various parts and organs of the body. These terms should be learned early on as they will be used in lecture, lab, and

Illustration by Jamey Garbett. © 2003 Mark Nielsen.

on tests. Read through the terms below and learn them for the first week of lecture. Note that following each description is an example of how the term can be used in a sentence.

Anatomical Position

It is important to have a reference point for descriptive purposes. The anatomical position is the reference point. It is the position from which anatomical structures are described. In the anatomical position, the body is erect, the upper limbs are at the sides, and the palms face forward. The lower limbs are together. The figure on this page is in the anatomical position. Realize that while this position is the basis of descriptive anatomy, it is not the natural position your body assumes when stand up in a relaxed natural position.

Dorsal

Dorsum is Latin for back and pertains to the back surface of the body. Exceptions to this are the dorsal surface of the foot which is the top aspect of the foot, and the dorsal surface of the penis which is the surface facing forward when the penis is flaccid. These uses of dorsal come from our ancestral quadrupedal stance, when the top surface of the foot and the front side of the penis did face dorsally.

Example: Touch the bones of your vertebral column on the dorsal surface of the trunk.

Ventral

Venter is Latin for belly and pertains to the belly surface of the body. In the case of the hand and foot the ventral surfaces are called the palmar and plantar surfaces, respectively.

Example: The umbilicus or belly button is located on the ventral surface of the body.

Posterior

Similar to dorsal in human anatomy, referencing the back side of the body or one of its parts.

Example: The vertebral column lies posterior to the heart.

Anterior

Similar to ventral in human anatomy, referencing the front side of the body.

Example: The knee caps form the anterior surface of the knee.

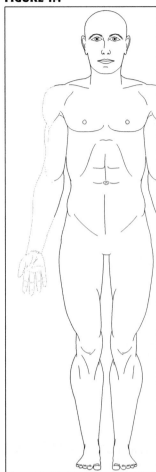

FIGURE 1.4

Illustration by Jamey Garbett. © 2003 Mark Nielsen.

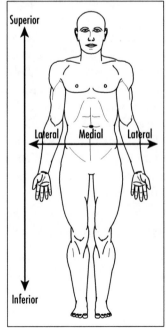

FIGURE 1.5
Anterior or ventral view

Illustration by Jamey Garbett. © 2003 Mark Nielsen.

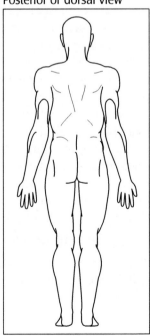

FIGURE 1.6
Posterior or dorsal view

Illustration by Jamey Garbett. © 2003 Mark Nielsen.

Superior

A term meaning above. Can be used, in human anatomy, as a substitute for cranial/cephalic. See cranial below.

Example: The head is superior to the neck.

Inferior

A term meaning below. Can sometimes be used instead of caudal but not in all instances. See caudal below.

Example: The abdomen is inferior to the chest.

Medial

Relating to the middle or center. Towards the median or mid sagittal plane. (See definitions of these planes on the following pages.)

Example: The nose is medial to the ears.

Lateral

From the Latin word latus meaning side. Towards the side of the structure or body. Farther from the median plane.

Example: The eyes are lateral to the body's midline or the eyes are lateral to the nose.

Proximal

From the Latin proximus meaning nearest. A term used to reference landmarks on structures that are branches or appendages of other structures (e.g., limbs, bronchial tree, nerves, etc.), meaning nearer the base or point of origin of the structure. This is useful terminology because it does not lose its meaning as the limb changes position as it is moved.

Example: The shoulder joint is proximal to the elbow.

Distal

From the Latin distantia meaning remote or far. A term used to reference landmarks on structures that are branches or appendages of other structures (e.g., limbs, bronchial tree, etc.), meaning farther from the base or point of origin of the structure. An exception to this is its use in dentistry. Here distal refers to the side of the tooth toward the back of the mouth. The term mesial is used to refer to the other side of the tooth, the side toward the midline (the point between the two first incisors) of the dental arch or jaw.

Example: The wrist is distal to the elbow.

Cranial

From the Greek word kranion meaning skull. Toward or related to the head end of the body. Another term used to describe the head is cephalic.

Example: The cervical vertebrae are cranial to the thoracic vertebrae.

FIGURE 1.7 Lateral view

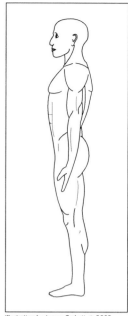

Illustration by Jamey Garbett. © 2003 Mark Nielsen.

FIGURE 1.8

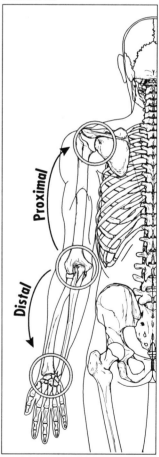

Illustration by Jamey Garbett. © 2003 Mark Nielsen.

Rostral

From the Latin word rostrum meaning beak. Toward the beak or snout end of the body. More commonly used in quadrupedal anatomy. In human anatomy it is used when describing relations within the brain. Rostral structures are towards the front of the head.

Example: The nose is the rostral most end of the body. The eyes are rostral outgrowths of the brain.

Caudal

From the Latin word cauda meaning tail. This term denotes the tail end of the body. This is properly used when referring to points on the trunk. The tail is a projection from the bottom of the trunk, therefore proper use of this term does not apply to points on the lower limbs below the tail. For this reason, it is not technically interchangeable with the term inferior.

Example: The tip of the tailbone is the caudal most point on our body. The lumbar vertebrae are caudal to the thoracic vertebrae.

Superficial

From the Latin word superficies meaning surface. This term denotes something on the surface or situated near the surface.

Example: The skin is the superficial-most organ of the human body.

Deep/Profundus

Situated at a deeper level in relation to other structures, that is, being beneath other layers of anatomy.

Example: The heart and lungs lie deep to the rib cage. The dermis is deep to the epidermis.

External

Meaning toward the outside of something or on the exterior of an organ or cavity. Toward the outside surface of a structure or body.

Example: We have hairs on the external surface of our body.

Internal

Meaning toward the inside or in the interior of an organ or cavity. Toward the inside surface of a structure or body.

Example: An ulcer is a pathological condition that begins on the internal surface of the digestive system organs.

FIGURE 1.9

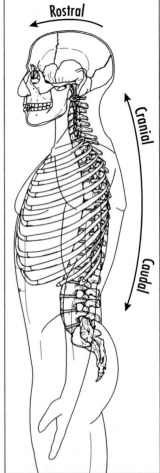

Illustration by Jamey Garbett. © 2003 Mark Nielsen.

FIGURE 1.10

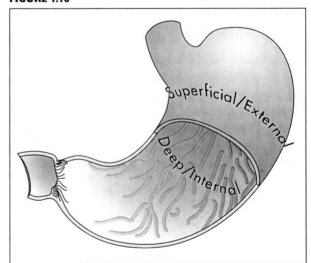

Illustration by Jamey Garbett. © 2003 Mark Nielsen.

Ipsilateral

From the Latin words ipsa meaning same and latus meaning side. Therefore, on the same side of the body.

Example: The right upper and lower limbs are on ipsilateral sides of the body. The ipsilateral limbs both moved forward at the same time.

Contralateral

From the Latin words contra meaning opposite and latus meaning side. Therefore, on opposite sides of the body.

Example: The right upper limb and left lower limb are on contralateral sides of the body. When you run, the contralateral upper and lower limbs move forward at the same time.

Anatomical Planes and Sections

In anatomical studies it is common for anatomists to cut the body into sections to better visualize and study deep or internal anatomy. The terms below describe the different planes and sections encountered in anatomical studies as well as in many medical procedures.

Planes

Imaginary planes oriented in three-dimensional space.

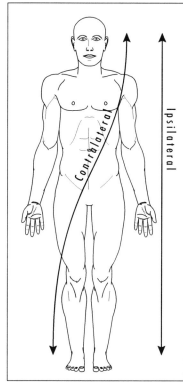

FIGURE 1.11

Sagittal Plane

An imaginary vertical plane that separates the body into right and left parts. Sagittal planes get their name from the fact that they run parallel to the sagittal suture of the skull (the anterior to posterior oriented suture on the top of the skull).

Median or Median Sagittal Plane

The imaginary vertical plane that divides the body longitudinally into exact right and left halves.

Paramedian or Parasagittal Plane

Any imaginary sagittal plane that is parallel to the median plane and divides the body into unequal right and left parts.

FIGURE 1.12

FIGURE 1.13

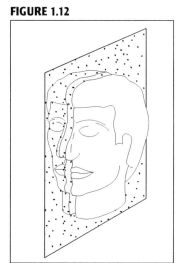

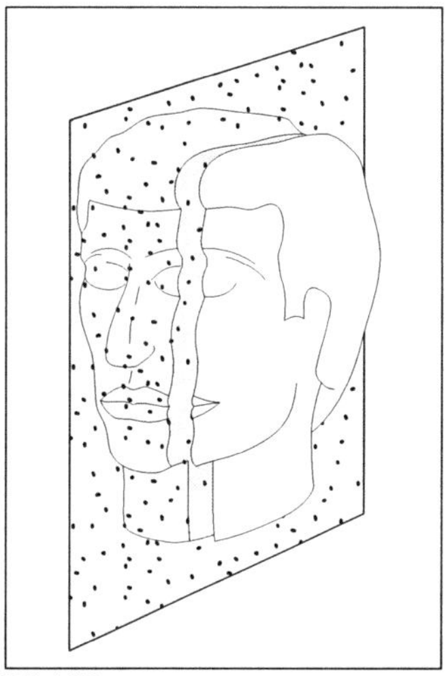

Coronal/Frontal Plane

These are imaginary vertical planes which course at right angles to the sagittal plane. They divide the body into anterior and posterior parts. The coronal suture of the skull follows this plane.

Horizontal Plane

This is the imaginary plane that parallels the horizon. A plane situated at right angles to either of the vertical planes of space.

FIGURE 1.14

FIGURE 1.15

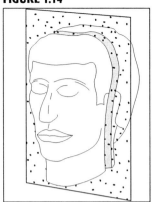
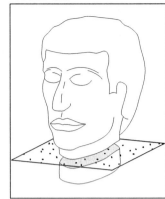

Illustrations by Jamey Garbett. © 2003 Mark Nielsen.

Sections

Terms describing actual cuts or slices of the body or its parts.

Longitudinal Sections

Sections, slices, or cuts of the body which run in the direction of the long axis of the body. They may be made in the median, paramedian, or coronal planes.

Transverse Sections

Sections of the body which are made at right angles to the longitudinal axis of the body or its parts.

Parts of the Body

There are a number of terms anatomists use to describe the various regions or parts of the body. These terms have specific meanings and are often misused by the layperson. It is important to understand the correct usage of the terms as outlined below.

Head

The head is the superior most portion of the human body. The head consists of the boney cranium and the soft tissues that surround it and are within it. The cranium, or skull, has two major regions, the cranial vault that houses the brain and the facial skeleton and its associated soft tissues that form the eyes, nose, cheeks, and jaw apparatus.

Trunk

The trunk is the portion of the body that incorporates the vertebral column, in other words, the body excluding the head and extremities. The trunk is divided into four regions: the neck, thorax, abdomen, and pelvis. The neck corresponds to the seven cervical vertebrae, the thorax to the twelve thoracic vertebrae, the abdomen to the five lumbar vertebrae, and the pelvis to the sacrum and coccyx.

FIGURE 1.16

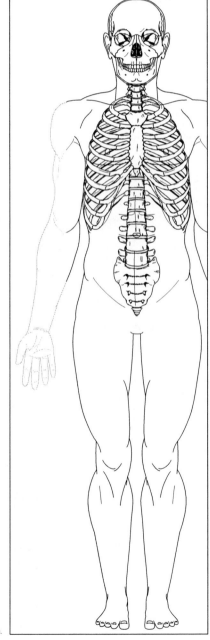

Illustration by Jamey Garbett. © 2003 Mark Nielsen.

Upper Limb

FIGURE 1.17

The upper limb is the entire structure from the pectoral girdle to the ends of the fingers.

Pectoral Girdle (Shoulder Girdle)

The pectoral girdle is the support structure at the base of the upper limb. It consists of the scapula and clavicle.

Brachium (Arm)

The brachium is that portion of the upper limb between the shoulder and the elbow. Its skeletal element is the humerus.

Antebrachium (Forearm)

The antebrachium is that portion of the upper limb between the elbow and the wrist. Its skeletal elements are the ulna and radius.

Manus (Hand)

The manus is the portion of the upper limb that includes the wrist bones, the metacarpal bones of the palm, and to the bones of the fingers.

Carpus (Wrist)

The carpus, or wrist, is the proximal most portion of the hand. A narrow region comprised of the eight small carpal bones, which join the longer metacarpal bones of the palm with the radius of the antebrachium.

Metacarpus

The metacarpus is the region forming the palm and dorsum of the hand. It consists of the five metacarpal bones.

Digits (Fingers and Thumb)

The digits are the five ray-like extensions at the distal end of the hand that are commonly referred to as the fingers and thumb. The phalangeal bones form the skeletal structure of the digits.

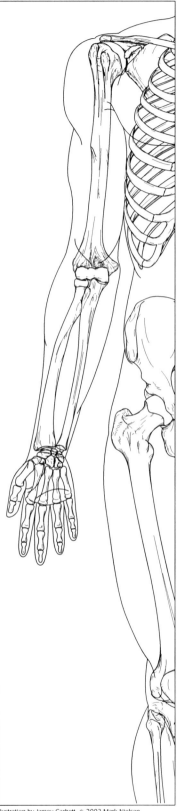

Illustration by Jamey Garbett. © 2003 Mark Nielsen.

Lower Limb

The lower limb is the entire structure from the pelvic girdle to the ends of the toes.

Pelvic Girdle

The pelvic girdle is the skeletal base of the inferior limb, or os coxa bone. This bone forms from the developmental fusion of three bones, the ilium, ischium, and pubis. The right and left pelvic girdles unite with the sacral region of the vertebral column to anchor the lower limb to the axial skeleton of the trunk.

Clunes (Buttocks)

The cluneal region, or buttocks, is the fleshy muscles and fat that cover the posterior and lateral aspect of the pelvic girdle.

Coxa (Hip)

The hip is the junction, or joint, formed by the femur, bone of the thigh, and os coxa at the proximal end of the inferior limb.

Thigh

The thigh is that portion of the lower limb between the hip and the knee. The femur is the skeletal element of the thigh.

Crus (Leg)

The crus, or crural region, is that portion of the lower limb between the knee and the ankle. Its skeletal elements are the tibia and fibula.

Pes (Foot)

The pes is that portion of the lower limb called the foot. It consists of three parts — the ankle at the proximal end, the metatarsus forming the middle portion of the foot, and the digits, or toes.

Tarsus (Ankle)

The tarsus is the heel end, or root of the foot. Its skeletal elements are the seven tarsal bones, which connect the tibia and fibula to the metatarsals.

Metatarsus

The metatarsus is the intermediate portion of the foot. The five slender metatarsal bones are its skeletal elements.

Digits (Toes)

The digits are the five ray-like extensions at the distal end of the foot commonly referred to as the toes. The phalangeal bones are the skeletal elements of the toes.

FIGURE 1.18

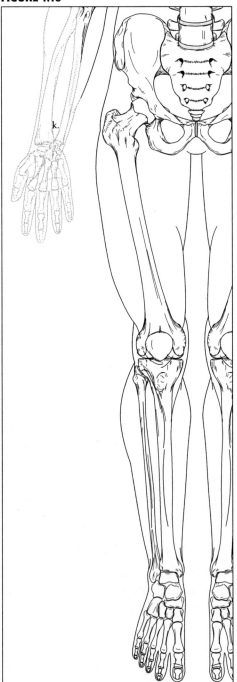

Illustration by Jamey Garbett. © 2003 Mark Nielsen.

Cavities

Within some of the regions of the body previously described, there are hollow cavities that house, or contain, organs that are internal to the bony or muscular walls of the cavities. There are three major body cavities.

Cranial Cavity

The cranial cavity is the cavity within the bones of the skull. This cavity houses the brain and its associated structures.

Thoracic Cavity

The thoracic cavity is the cavity within the musculoskeletal wall of the rib cage. This cavity houses the heart, lungs, and other thoracic organs.

Abdominopelvic Cavity

The abdominopelvic cavity is the cavity enclosed by the abdominal muscles and pelvic bones and muscles. It is separated from the thoracic cavity by the muscular diaphragm. This cavity houses the organs of the abdomen and pelvis.

FIGURE 1.19

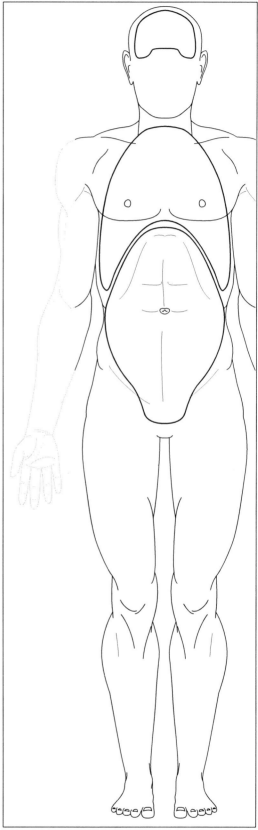

Illustration by Jamey Garbett. © 2003 Mark Nielsen.

Anatomical Nomenclature

1. In the illustration right, the human
    body is divided by three planes. Draw
    an illustration of the body as it
    appears sectioned in each of the three
    planes. Only an outline is necessary,
    but be sure to identify your three
    drawings.

FIGURE 1.20 The three planes of the body

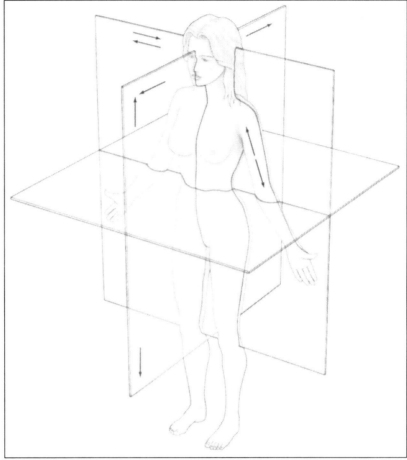

From *Laboratory Guide for Human Anatomy* by William J. Radke, copyright © 2002 John Wiley & Sons, Inc. Reprinted by permission of John Wiley & Sons, Inc.

2. Complete the following sentences using correct anatomical language:

The head is ___a___ to the neck, and the neck is ___b___ to the head.

The skin of the brachium is ___c___ to its muscles, and the bones are ___d___ to the muscles.

The ___e___ and ___f___ are distal to the knee, while the ___g___ is proximal to the knee.

The scapula and the clavicles form the ___h___ .

The ilium, ischium, and pubis form the ___i___ girdle.

When walking, the ___j___ arm and leg swing forward together.

a. _____   e. _____   h. _____

b. _____   f. _____   i. _____

c. _____   g. _____   j. _____

d. _____

3. Label the illustration below with terminology from this chapter using colored pencils or brackets to indicate the regions:

Brachium	Digits	Pes
Antebrachium	Clunes	Tarsus
Manus	Coxa	Metatarsus
Carpus	Thigh (upper leg)	Digits
Metacarpus	Crus (leg)	

FIGURE 1.21 Male and female in anatomical position

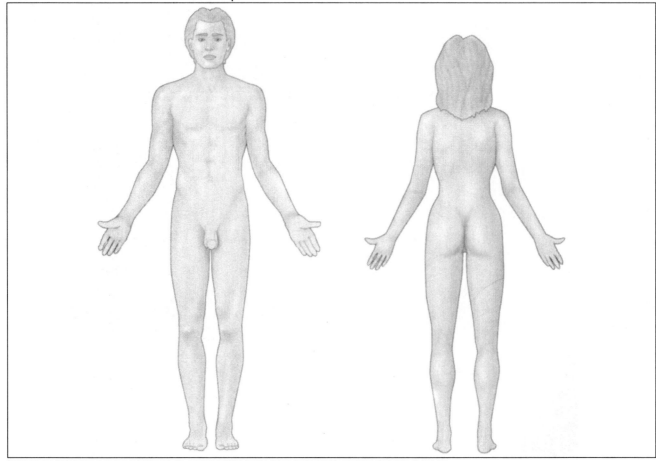

4. Label the diagram below, illustrating the cavities listed. Using your textbook or another source, list the structures found in each cavity.

Dorsal Body Cavity:

a. Cranial cavity

b. Vertebral cavity

Ventral Body Cavity:

a. Abdominopelvic cavity

    (1) Abdominal cavity

    (2) Pelvic cavity

b. Thoracic cavity

    (1) Pleural cavities

    (2) Pericardial cavity

FIGURE 1.22 Body cavities

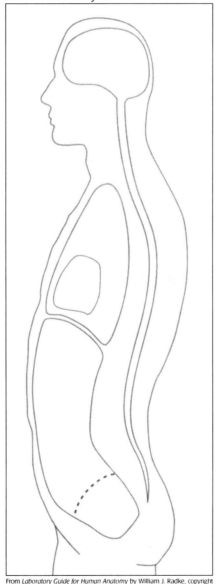

Axial Skeleton

The axial skeleton is composed of three regions: (1) the skull, (2) the vertebral column, and (3) the thoracic cage. The axial skeleton protects the central nervous system in the cranial and vertebral cavities, protects the heart and lungs, and facilitates ventilation of the lungs. The axial skeleton connects to the pectoral girdle at the sternoclavicular joint (the articulation of the sternum and clavicle); the weight of the body is transferred from the axial skeleton to the pelvic girdle at the sacroiliac joint (the articulation of the sacrum and the ilium of the pelvis).

OBJECTIVES

☑ Know the functions particular to the axial skeleton
☑ Identify the bones and major landmarks of the skull
☑ Identify the parts of a vertebra, and discern the differences between cervical, thoracic, lumbar, sacral, and coccygeal vertebrae
☑ Understand the nature of the articulations between bones of the axial skeleton and between the axial and appendicular skeleton

STRUCTURES YOU ARE RESPONSIBLE FOR IDENTIFYING

Cranial Bones
Ethmoid bone
Frontal bone
Maxillary bone
Inferior nasal conchae
Lacrimal bone
Mandible
Nasal bone
Occipital bone
Palatine bone
Parietal bone
Sphenoid bone
Temporal bone
Vomer
Zygomatic bone
Cranial Sutures
Coronal suture
Sagittal suture
Squamous suture
Lambdoid suture

Cranial Landmarks
Paranasal sinuses
Occipital condyle
Mastoid process

Styloid process
Foramen magnum
Optic canal
Sella turcia
Pterygoid process
Cribiform plate
External acoustic meatus
Internal acoustic meatus
Coronoid process (of mandible)
Ramus (of mandible)

Hyoid Bone
Vertebrae
Intervertebral disc
Body
Inferior articular process
Lamina
Pedicle
Vertebral arch
Spinous process
Superior articular process
Transverse process
Vertebral foramen
Intervertebral foramen

Cervical Vertebrae
Transverse foramen
Atlas
Axis
Dens (of Axis)

Thoracic Vertebrae
Costal facets

Sacrum (Sacral Vertebrae)
Anterior sacral foramina
Posterior sacral foramina
Sacral canal
Alae
Median sacral crest

Coccyx
Ribs
Head
Neck
Tubercle
Costal cartilages

Sternum
Manubrium
Body
Zyphoid process

Vertebral Column

The vertebral column consists of numerous serial elements of a segmental origin termed vertebrae, singular vertebra. There are 32 to 34 vertebrae and they are divided into five regions—cervical, thoracic, lumbar, sacral, and coccygeal.

General Characteristics of a Vertebra

The typical vertebra consists of the following parts. Identify each of these on various vertebrae.

FIGURE 2.1

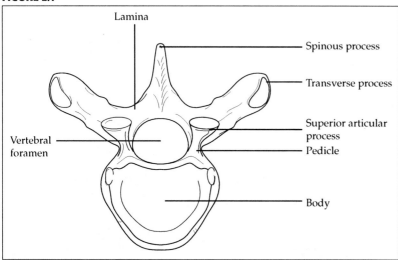

FIGURE 2.2

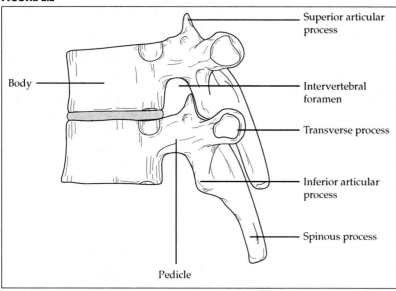

Illustrations by Jamey Garbett. © 2003 Mark Nielsen.

FIGURE 2.3

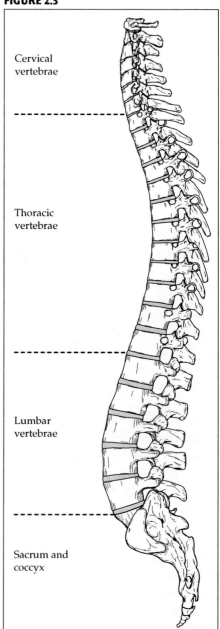

Cervical Vertebrae

The cervical vertebrae are seven in number and may be readily identified by the transverse foramen in their transverse processes. The first two cervical vertebrae are highly modified. The first cervical vertebra, the atlas, derives its name from the fact that it supports the globe-like head, just as the Greek god Atlas supported the globe-like world. It is annular in shape and unlike all other vertebrae it lacks a body. Examine the atlas and note its similarities to and differences from other cervical vertebrae. The second cervical vertebra, the axis, is the pivot on which the atlas rotates. It has a characteristic toothlike projection, the dens or odontoid process, projecting vertically from its body. Compare the axis to a typical cervical vertebrae.

FIGURE 2.4 Atlas, superior view

Illustrations by Jamey Garbett. © 2003 Mark Nielsen.

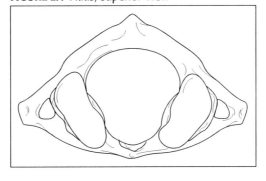

FIGURE 2.5 Axis, anterior view

FIGURE 2.6 Axis, lateral view

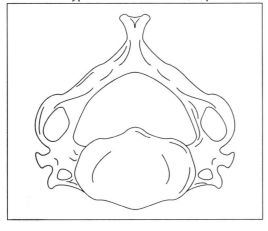
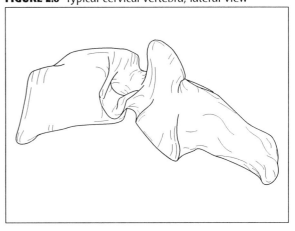

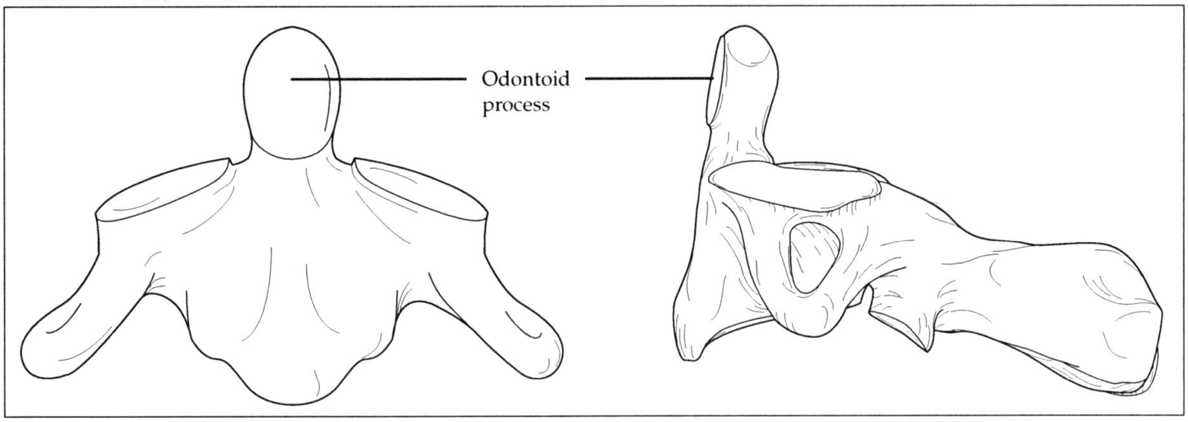

Odontoid process

FIGURE 2.7 Typical cervical vertebra, superior view

FIGURE 2.8 Typical cervical vertebra, lateral view

Thoracic Vertebrae

The thoracic vertebrae are twelve in number and are easily identified by the articular surfaces called costal facets. The costal facets are the articular surfaces the thoracic vertebrae form with the ribs. Notice the locations and number of these costal facets on a typical thoracic vertebra. Also, notice that the spinous processes of a thoracic vertebra tend to slant down and back and the transverse processes are strong club-like projections of bone.

FIGURE 2.9 Thoracic vertebra, superior view

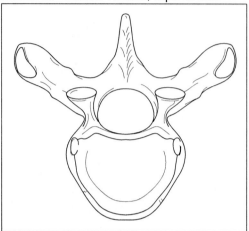

FIGURE 2.10 Thoracic vertebra, lateral view

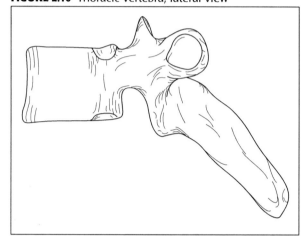

FIGURE 2.11 Thoracic vertebra, anterior view

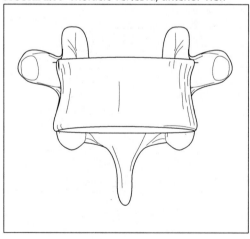

FIGURE 2.12 Thoracic vertebra, posterior view

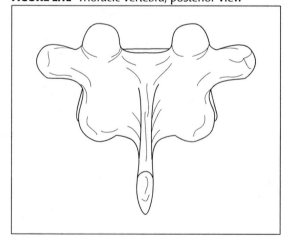

Illustrations by Jamey Garbett. © 2003 Mark Nielsen.

Articular Surfaces on a Thoracic Vertebra

An articular facet is a flattened synovial articular surface on a bone. A process is the projection of bone upon which the facet is located. Can you differentiate between the superior and inferior articular processes and the superior and inferior articular facets?

Costovertebral Articulations

FIGURE 2.13

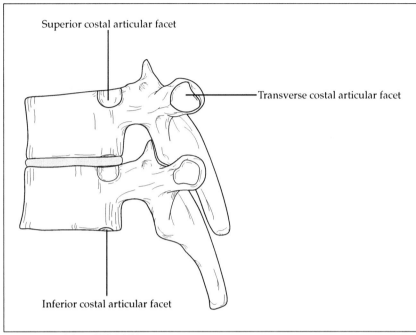

Superior costal articular facet

Transverse costal articular facet

Inferior costal articular facet

Illustration by Jamey Garbett. © 2003 Mark Nielsen.

Intervertebral Articulations

FIGURE 2.14

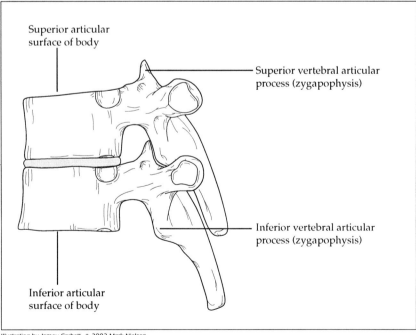

Superior articular surface of body

Superior vertebral articular process (zygapophysis)

Inferior vertebral articular process (zygapophysis)

Inferior articular surface of body

Illustration by Jamey Garbett. © 2003 Mark Nielsen.

Lumbar Vertebrae

The lumbar vertebrae are typically large, bulky vertebrae that can be distinguished by the absence of transverse foramina and costal facets. There are five lumbar vertebrae. Notice the orientation of the spinous process and compare it to the thoracic spinous process.

FIGURE 2.15 Lumbar vertebra, superior view

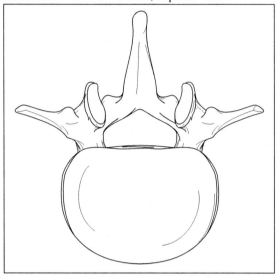

FIGURE 2.16 Lumbar vertebra, lateral view

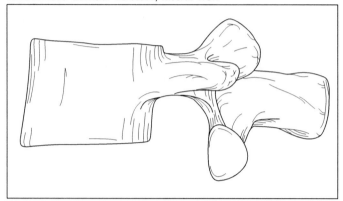

FIGURE 2.17 Lumbar vertebra, anterior view

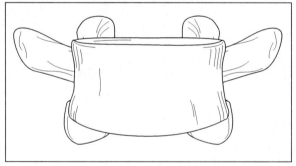

FIGURE 2.18 Lumbar vertebra, posterior view

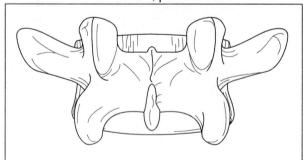

Sacrum

The sacrum is composed of the five fused sacral vertebrae. It is a large, triangular element positioned below the lumbar region where it forms the posterior pelvic wall. It combines with the os coxae of each side to form the pelvis.

Coccyx

The coccyx is the small triangular tail at the caudal end of the sacrum. It forms from the fusion of three to five vertebrae, most typically four. Identify the coccyx on the articulated skeleton in the lab.

FIGURE 2.19 Sacrum and coccyx

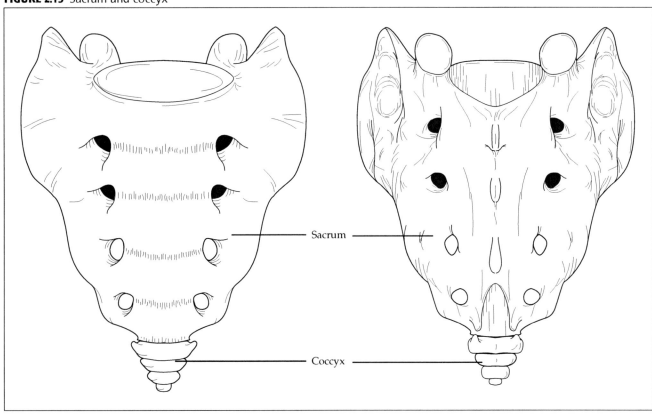

(a) anterior view   (b) posterior view

Illustration by Jamey Garbett. © 2003 Mark Nielsen.

Rib Cage

The ribs and associated sternum form a protective cage around the thoracic viscera. Not only do they protect these important viscera, but they also play an important mechanical role in the ventilation of the lungs.

Sternum

The sternum consists of three parts:

Manubrium

This is the handle-like cranial portion of the sternum.

Body

The body is the large central portion of the sternum and it consists of four fused segments or sternebrae.

Xiphoid Process

This sword-like process is located at the caudal end of the sternum.

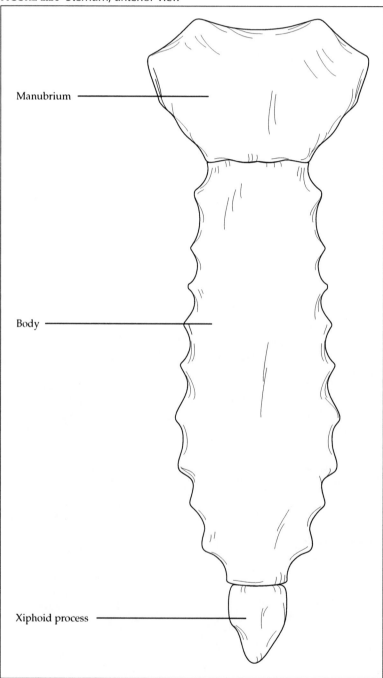

FIGURE 2.20 Sternum, anterior view

Manubrium

Body

Xiphoid process

Illustration by Jamey Garbett. © 2003 Mark Nielsen.

Ribs

There are typically twelve paired ribs. They form arches of bone that join the vertebral column **posteriorly** with the sternum anteriorly. They join with the sternum via the costal cartilages. Identify the following **anatomy** on a typical rib:

Head

This is the rounded posterior extremity that articulates with the vertebral column.

Neck

This is the slightly constricted region of bone adjacent to the head.

Tubercle

This is a small projection of bone at the junction of the neck and the shaft of the rib. The tubercle has an articular facet for its articulation with the transverse process of the thoracic vertebrae.

Shaft

This is the thin, flattened remainder of the rib that makes of the greatest part of the bone.

Costal Groove

This is a shallow groove on the inferior, internal surface of the rib. The intercostal artery, vein, and nerve run within this groove as they course around the rib cage.

FIGURE 2.21 Typical rib, anterior view

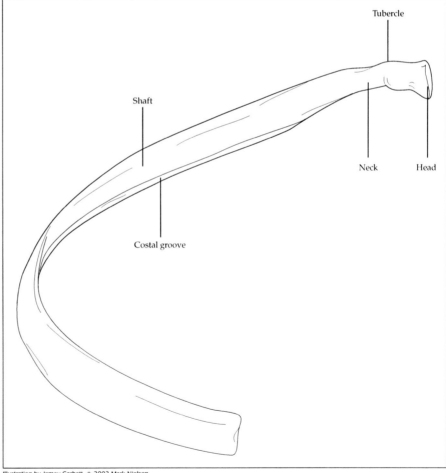

Illustration by Jamey Garbett. © 2003 Mark Nielsen.

Classification of the Ribs

Ribs are classified into three categories:

True Ribs

The first seven ribs are termed true ribs because their costal cartilage is connected directly to the sternum.

False Ribs

The next five ribs are called false ribs because their costal cartilage does not join the sternum directly, instead it is joined to the costal cartilage of the more superior ribs or is not attached at all.

Floating Ribs

Ribs 11 and 12, and in some races rib 10, have small costal cartilages that are not attached anteriorly. Because they do not attach to the other ribs, they are termed floating ribs.

FIGURE 2.22 Rib cage, lateral view

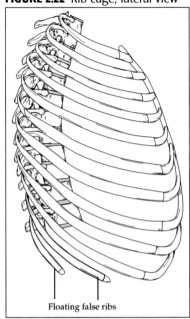

Floating false ribs

Illustration by Jamey Garbett. © 2003 Mark Nielsen.

FIGURE 2.23 Rib cage, anterior view

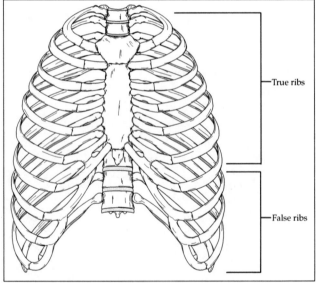

True ribs

False ribs

Illustration by Jamey Garbett. © 2003 Mark Nielsen.

Hyoid Bone

The hyoid bone is a U-shaped bone located in the anterior portion of the neck just inferior to the mandible. It serves as an important attachment site for the muscles of the tongue and certain ventral body wall muscles of the neck.

FIGURE 2.24 Hyoid bone, lateral view

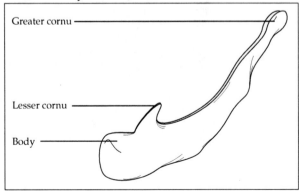

Greater cornu

Lesser cornu

Body

Illustration by Jamey Garbett. © 2003 Mark Nielsen.

FIGURE 2.25 Hyoid bone, anterior view

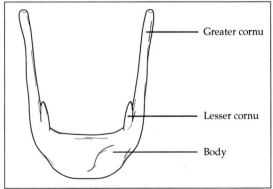

Greater cornu

Lesser cornu

Body

Illustration by Jamey Garbett. © 2003 Mark Nielsen.

Cranium

The cranium is a composite structure comprised of 29 bones, when counting the ear ossicles and hyoid bone. The bones of the cranium range from simple, nondescript plates of bone to the most intricate, complex bones of the skeleton. The bones of the cranium have a wide range of important functions. These functions, to name a few, include protecting the delicate brain tissue, fixing the vestibular apparatus of the inner ear in three dimensional space, maintaining open air passageways for respiration, and acquiring and processing food. The cranium consists of two main regions. One region, the neurocranium or brain box, is the region that surrounds and encases the brain. The other region is the viscerocranium or facial skeleton. This is the area contributing to the orbits, nasal cavity, and oral cavity.

FIGURE 2.26 Cranium, lateral view with neurocranium shaded

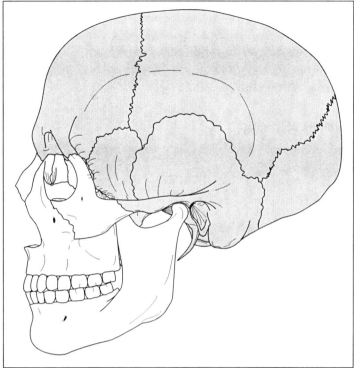

Illustration by Jamey Garbett. © 2003 Mark Nielsen.

FIGURE 2.27 Cranium, lateral view with viscerocranium shaded

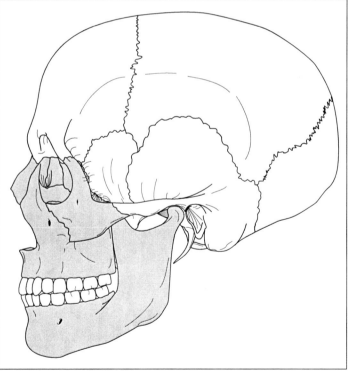

Illustration by Jamey Garbett. © 2003 Mark Nielsen.

General Features of the Cranium
Cranial Cavity

The cranial cavity is the housing for the brain. The shape of the cavity conforms with the shape of the brain. Note the deep impressions on the inside surfaces of the cavity. These are impressions left from large venous sinuses that drain blood from the brain. You can also identify impressions made by some of the arteries found in the connective tissue covering of the brain.

Foramen Magnum

The foramen magnum is the large opening in the base of the cranial vault through which the spinal cord passes as it exits the cranium.

Orbit

The orbit is a conical depression that contains the eyeball and the extrinsic muscles that produce movements of the eye. Can you list all the bones that contribute to the wall of the orbit?

Nasal Cavity

The nasal cavity is the initial passageway of the respiratory tract. It is bordered by many delicate bones that play an important role in increasing the surface area of the respiratory epithelium.

Osseous Sinuses

The osseous sinuses are cavitations within the cranium bones that form connections with the nasal cavity. Examine the sectioned craniums in the lab and identify the three large sinuses of the frontal, maxillary, and sphenoid bones.

Oral Cavity

The oral cavity is the region often referred to as the mouth. It is bounded above by the maxillary and palatine bones and below by the mandible. The teeth are a prominent feature of this region.

FIGURE 2.28

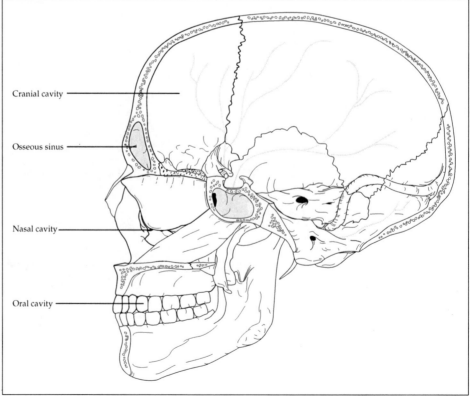

Cranial cavity

Osseous sinus

Nasal cavity

Oral cavity

Illustration by Jamey Garbett. © 2003 Mark Nielsen.

Bones of the Cranium

The following pages contain brief individual descriptions along with illustrations of each bone of the cranium. It is your responsibility to learn the name of each cranium bone and be able to identify each on a fully articulated cranium. The bone illustrations are presented in a manner (shaded grey) that illustrates each bone's position in the cranium from a number of perspectives.

Frontal Bone

The frontal bone has the shape of a shallow, irregular cap and forms the region known as the forehead. It forms the anterior wall of the cranial vault. Its inferior surface has a horizontal orbital plate that forms the greater part of the roof of the orbit.

FIGURE 2.29 Cranium, anterior view with frontal bone shaded

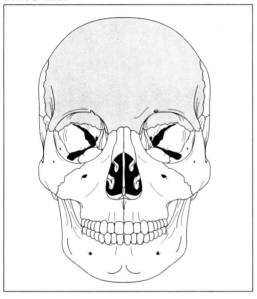

FIGURE 2.30 Cranium, superior view with frontal bone shaded

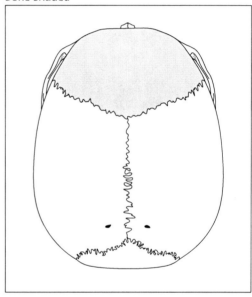

FIGURE 2.31 Cranium, lateral view with frontal bone shaded

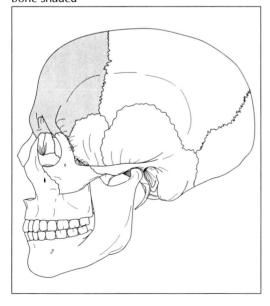

FIGURE 2.32 Cranium, sagittal view with frontal bone shaded

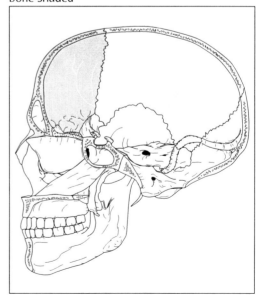

Parietal Bone

The parietal bone is an irregular four-sided bone with a gently arching shape. It forms much of the superior and lateral walls of the cranial vault.

FIGURE 2.33 Cranium, superior view with parietal bones shaded

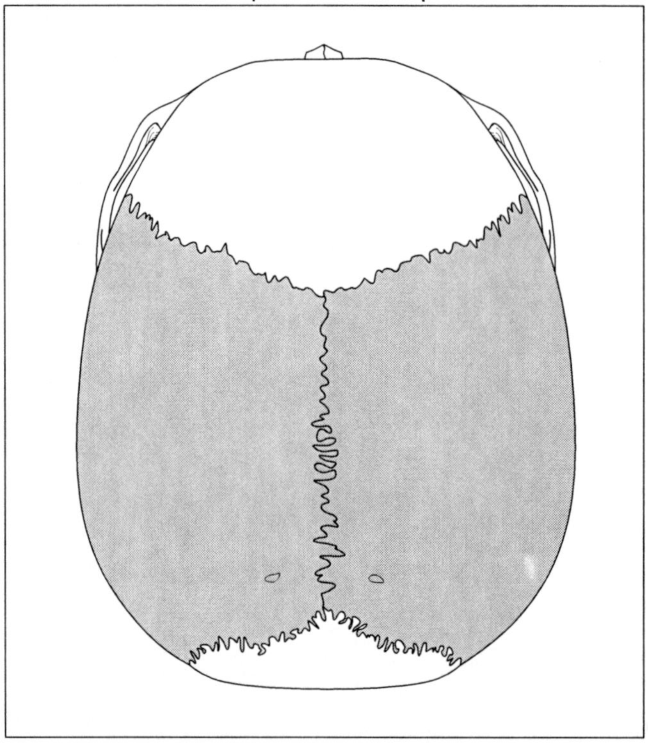

FIGURE 2.34 Cranium, lateral view with parietal bone shaded

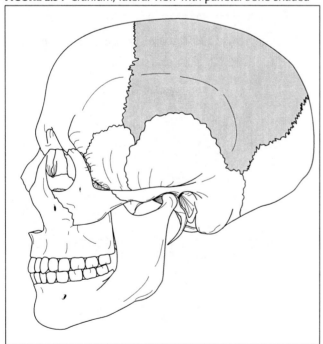

FIGURE 2.35 Cranium, sagittal view with parietal bone shaded

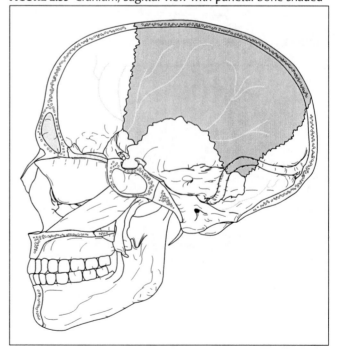

Occipital Bone

The occipital bone is a deeply concave bone with a trapezoid shape. It forms the posterior wall of the cranial vault and continues to the undersurface of the cranium where it surrounds the spinal cord to form the foramen magnum.

FIGURE 2.36 Cranium, inferior view with occipital bone shaded

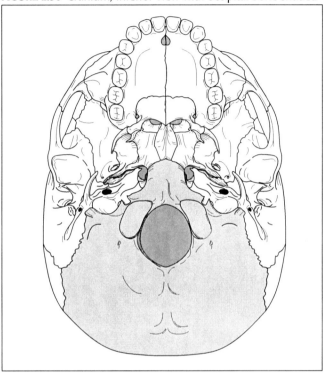

FIGURE 2.37 Cranium, lateral view with occipital bone shaded

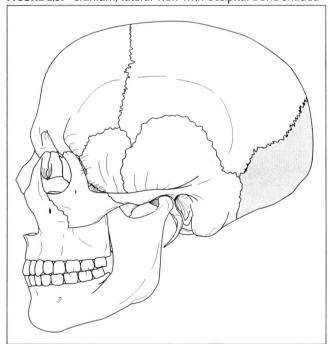

FIGURE 2.38 Cranium, sagittal view with occipital bone shaded

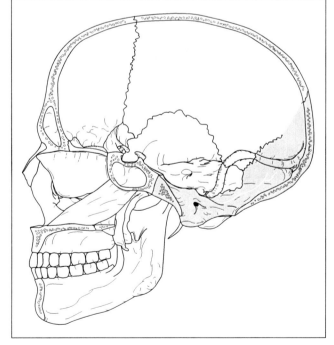

Illustrations by Jamey Garbett. © 2003 Mark Nielsen.

Temporal Bone

The temporal bone has an interesting developmental history as it arises from numerous centers of bone formation. You should be able to identify two major regions in the adult temporal bone. The flat squamous portion forms the sides of the cranial vault, while the rock-like petrous portion contributes to the base of the cranium and houses the middle and inner ear cavities. Within the middle ear cavity are the three small ear ossicles—the malleus, the incus, and the stapes.

FIGURE 2.39 Cranium, superior view with temporal bones shaded

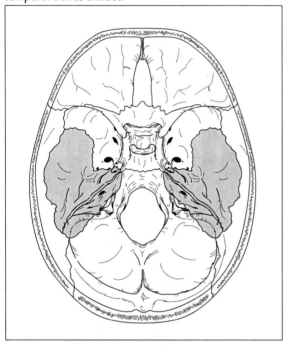

FIGURE 2.40 Cranium, inferior view with temporal bones shaded

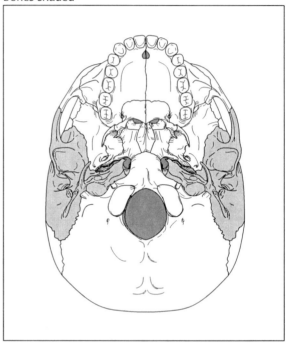

FIGURE 2.41 Cranium, lateral view with temporal bone shaded

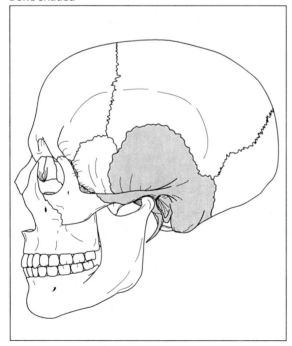

FIGURE 2.42 Cranium, sagittal view with temporal bone shaded

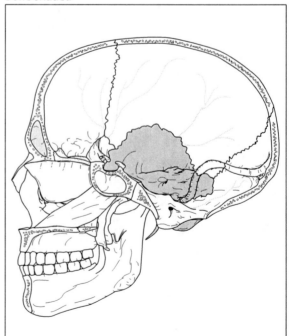

Maxillary Bone

The two irregular-shaped maxillary bones are the second largest bones of the facial portion of the cranium. Each maxillary bone articulates with all the other facial bones except the mandible. Together they form the entire upper jaw including the anterior roof of the mouth (hard palate), the floor and lateral walls of the nasal cavity, and the floor of the orbit.

FIGURE 2.43 Cranium, anterior view with maxillary bone shaded

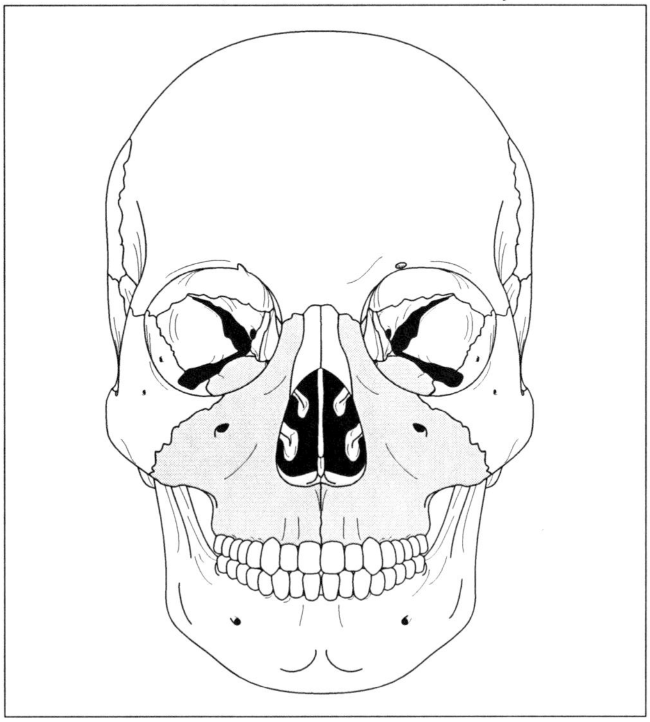

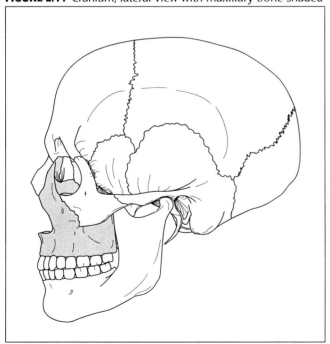

FIGURE 2.44 Cranium, lateral view with maxillary bone shaded

FIGURE 2.45 Cranium, inferior view with maxillary bone shaded

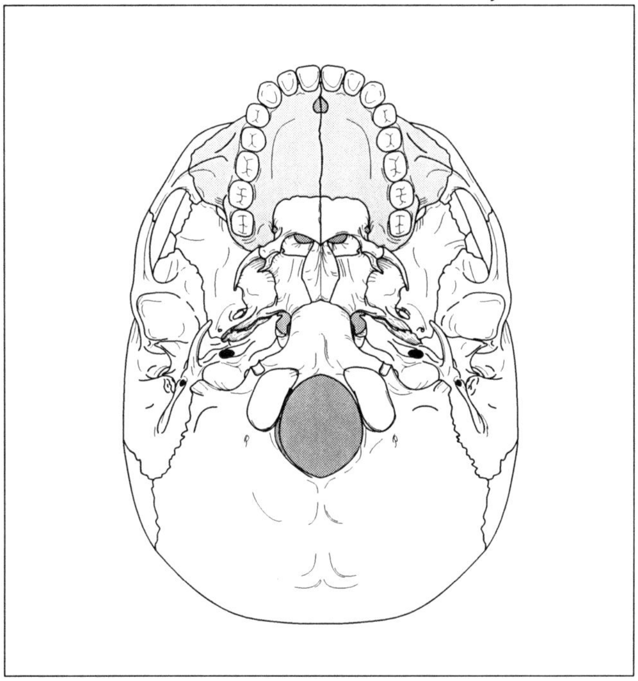

Mandible

The mandible is the bone of the lower jaw. It is the largest bone of the facial region. It has a U-shaped body that houses the lower tooth row. The posterior aspect of the body projects upward on each side as the mandibular ramus to form a synovial articulation with the base of the temporal bone.

FIGURE 2.46 Cranium, anterior view with mandible shaded

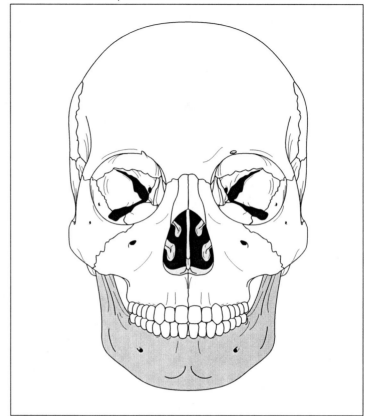

FIGURE 2.47 Cranium, lateral view with mandible shaded

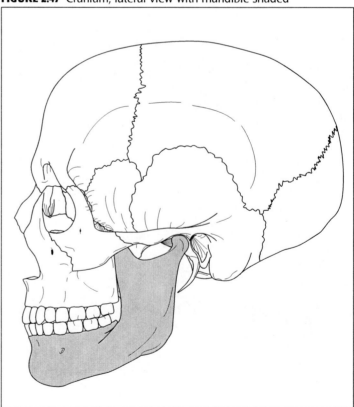

Illustrations by Jamey Garbett. © 2003 Mark Nielsen.

Zygomatic Bone

The zygomatic bone forms the prominence on the inferior and lateral border of the orbit referred to as the cheek. This bone contributes to the lateral wall and floor of the orbit. It forms a posterior projecting process that joins a similar process of the temporal bone to form the zygomatic arch.

FIGURE 2.48 Cranium, anterior view with zygomatic bones shaded

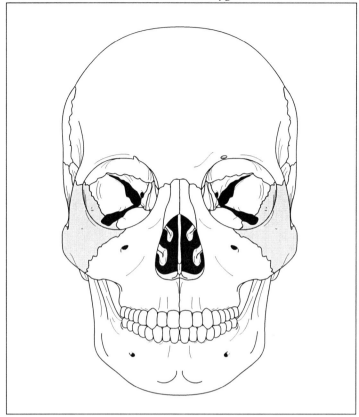

FIGURE 2.49 Cranium, inferior view with zygomatic bones shaded

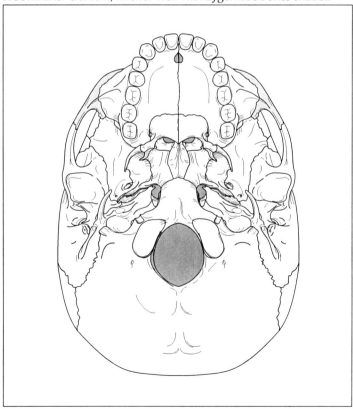

Nasal Bone

The nasal bone is an oblong bone of variable size that unites with its partner below the frontal bone to form the bridge of the nose and the anterior wall of the upper nasal cavity.

FIGURE 2.50 Cranium, anterior view with nasal bones shaded

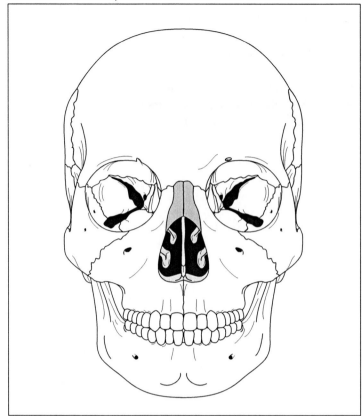

FIGURE 2.51 Cranium, lateral view with nasal bones shaded

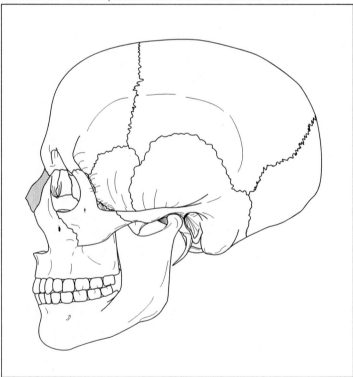

Illustrations by Jamey Garbett. © 2003 Mark Nielsen.

Sphenoid Bone

The sphenoid bone is a complex, irregular shaped bone that has the general appearance of a butterfly. It is wedged into the base of the cranium just in front of the temporal bones and the basilar part of the occipital bone. It contributes to the anterior and basal walls of the cranial vault, the posterior walls of the orbit, and the posterior aspect of the nasal cavity. It is visible from almost any internal or external view of the cranium.

FIGURE 2.52 Sphenoid bone, anterior view

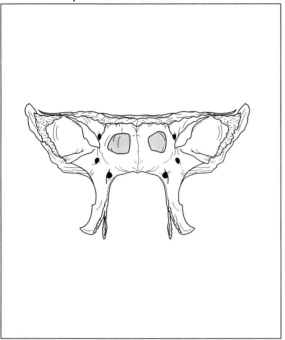

FIGURE 2.53 Cranium, anterior view with sphenoid bone shaded

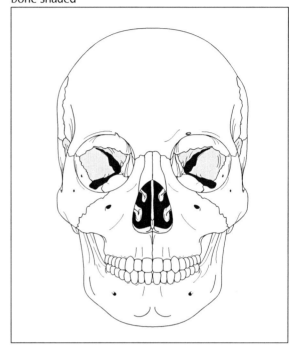

FIGURE 2.54 Cranium, superior view with sphenoid bone shaded

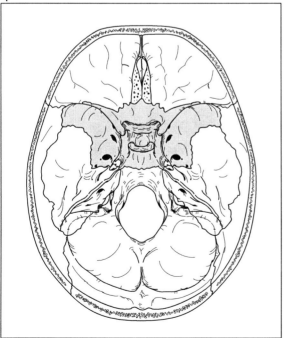

FIGURE 2.55 Cranium, inferior view with sphenoid bone shaded

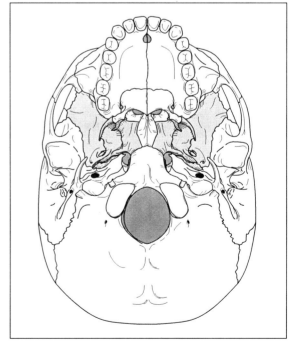

Sphenoid Bone

FIGURE 2.56 Dissected cranium, posterior view with sphenoid bone shaded

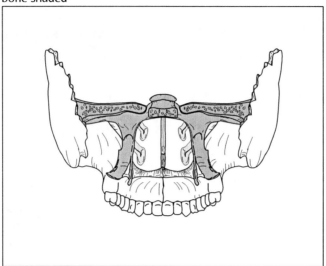

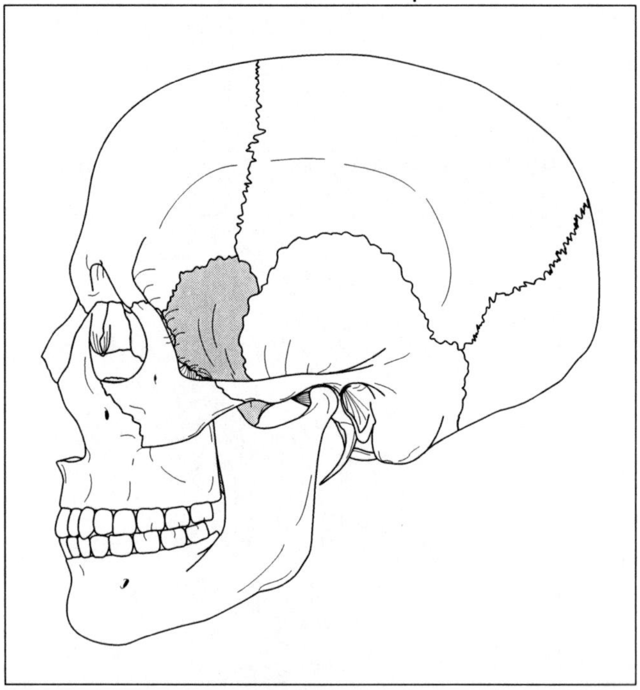

FIGURE 2.57 Cranium, lateral view with sphenoid bone shaded

FIGURE 2.58 Cranium, sagittal view with sphenoid bone shaded

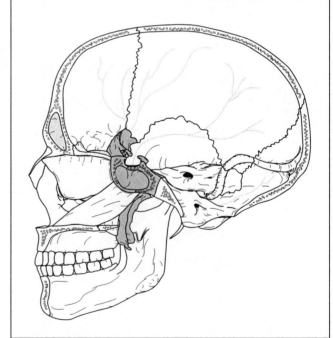

Illustrations by Jamey Garbett. © 2003 Mark Nielsen.

Ethmoid Bone

The ethmoid bone is a delicate, box-like bone that is wedged in between the frontal, maxillary, and sphenoid bones. It is located at the anterior part of the base of the cranium where it surrounds the olfactory nerves as they enter the nasal cavity. It forms the medial wall of the orbits, contributes to the bony septum of the nose, and forms the roof and upper lateral walls of the nasal cavity.

FIGURE 2.59 Cranium, anterior view with ethmoid bone shaded

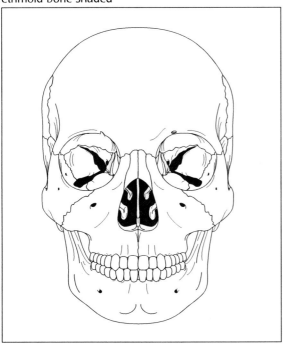

FIGURE 2.60 Cranium, superior view with ethmoid bone shaded

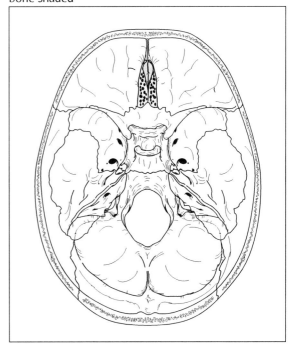

FIGURE 2.61 Cranium, sagittal view with ethmoid bone shaded

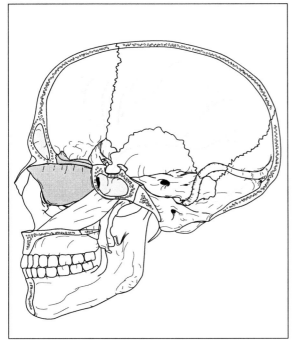

FIGURE 2.62 Cranium, parasagittal view with ethmoid bone shaded

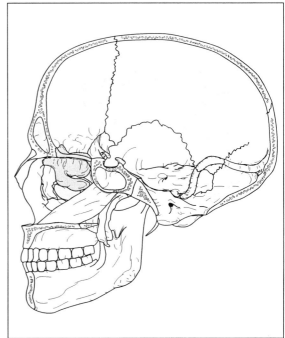

Lacrimal Bone

The lacrimal bone is a fragile, thin plate of bone that forms the anteromedial walls of the orbit. Notice the groove-like impression formed by the lacrimal sac and the passageway that carries the nasolacrimal or tear duct into the nasal cavity.

FIGURE 2.63 Cranium, anterior view with lacrimal bones shaded

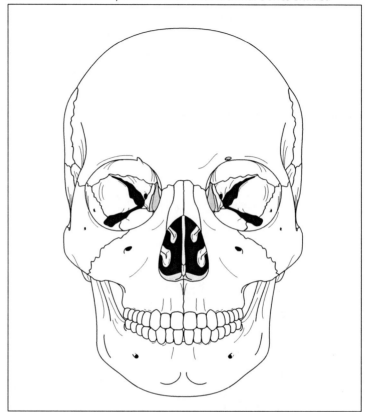

FIGURE 2.64 Cranium, lateral view with lacrimal bone shaded

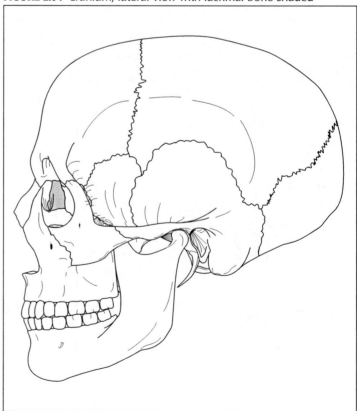

Illustrations by Jamey Garbett. © 2003 Mark Nielsen.

Palatine Bone

The small L-shaped palatine bones form the posterior third of the hard palate and contribute to the lateral walls of the nasal cavity.

FIGURE 2.65 Cranium, inferior view with palatine bones shaded

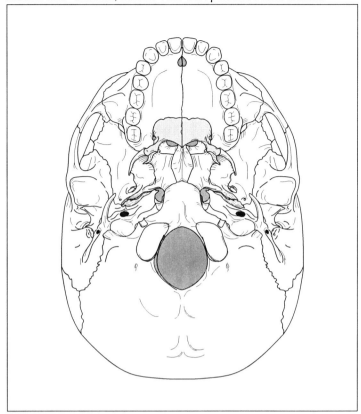

FIGURE 2.66 Cranium, lateral view with palatine bones shaded

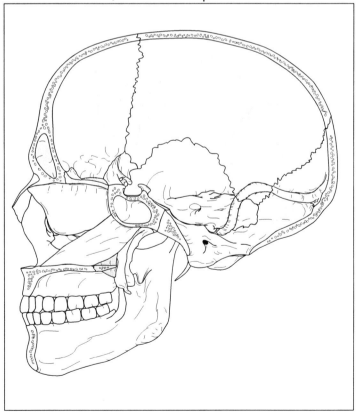

Vomer Bone

The vomer is a thin flat bone whose shape resembles a plow. It forms the posteroinferior part of the bony nasal septum.

FIGURE 2.67 Cranium, inferior view with vomer bone shaded

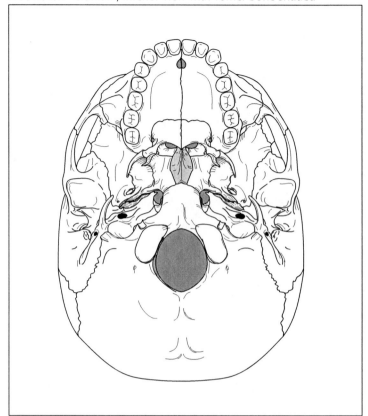

FIGURE 2.68 Cranium, sagittal view with vomer bone shaded

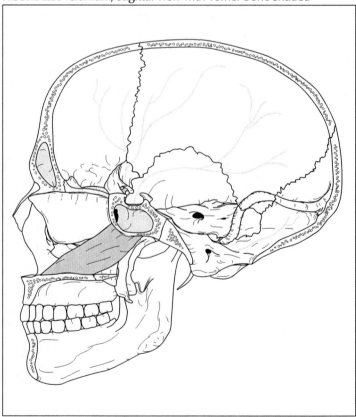

Inferior Nasal Concha Bone

The inferior nasal conchae are delicate, arching plates of bone that articulate with the medial surfaces of the maxillae. They project horizontally into the nasal cavity where they greatly increase the surface area of the nasal mucosa.

FIGURE 2.69 Cranium, anterior view with inferior nasal conchal bones shaded

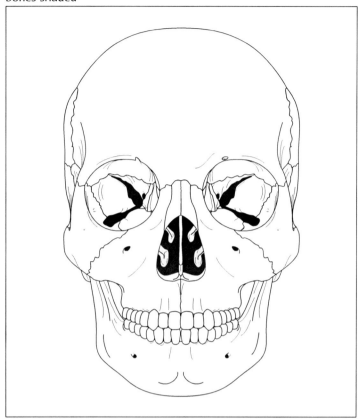

FIGURE 2.70 Cranium, sagittal view with inferior nasal conchal bone shaded

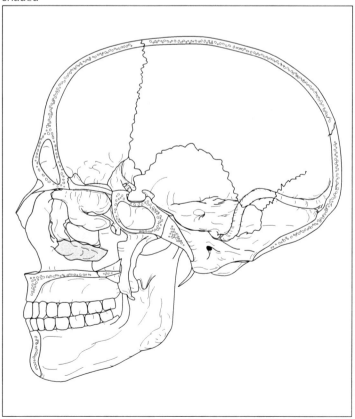

Illustrations by Jamey Garbett. © 2003 Mark Nielsen.

Axial Skeleton

Bone Identification Practice

Use the illustrations of the cranium on this page and the pages that follow and try to color code and label each of the bones of the cranium.

FIGURE 2.71

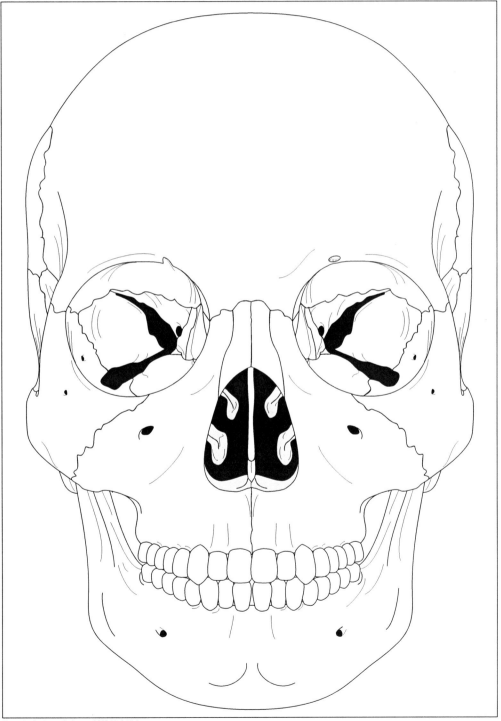

Illustration by Jamey Garbett. © 2003 Mark Nielsen.

FIGURE 2.72

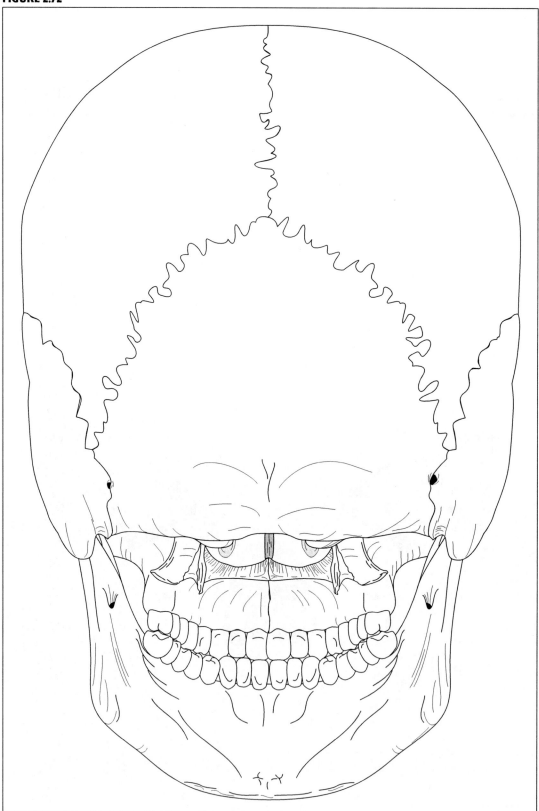

Illustration by Jamey Garbett. © 2003 Mark Nielsen.

FIGURE 2.73

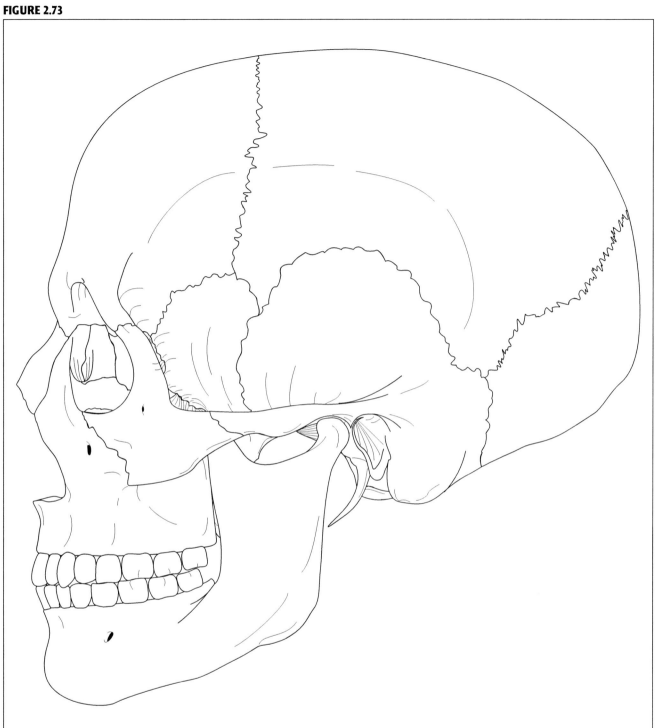

FIGURE 2.74

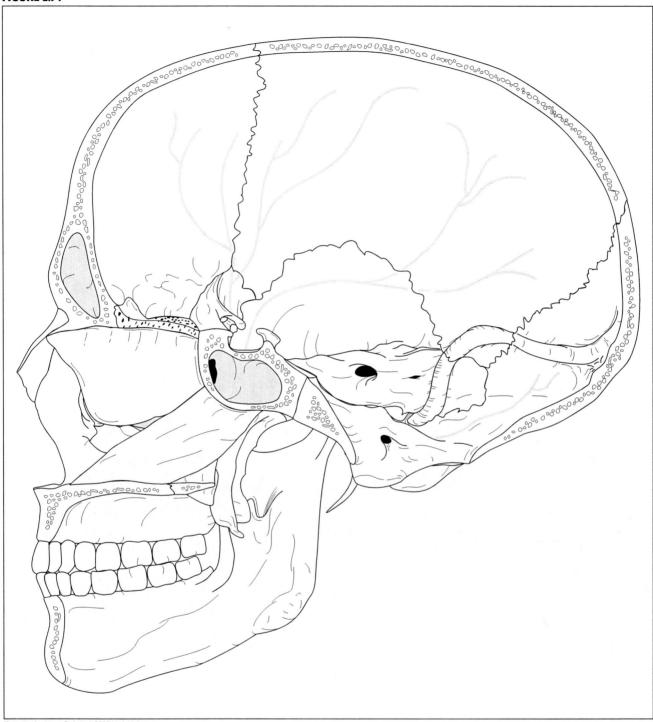

FIGURE 2.75

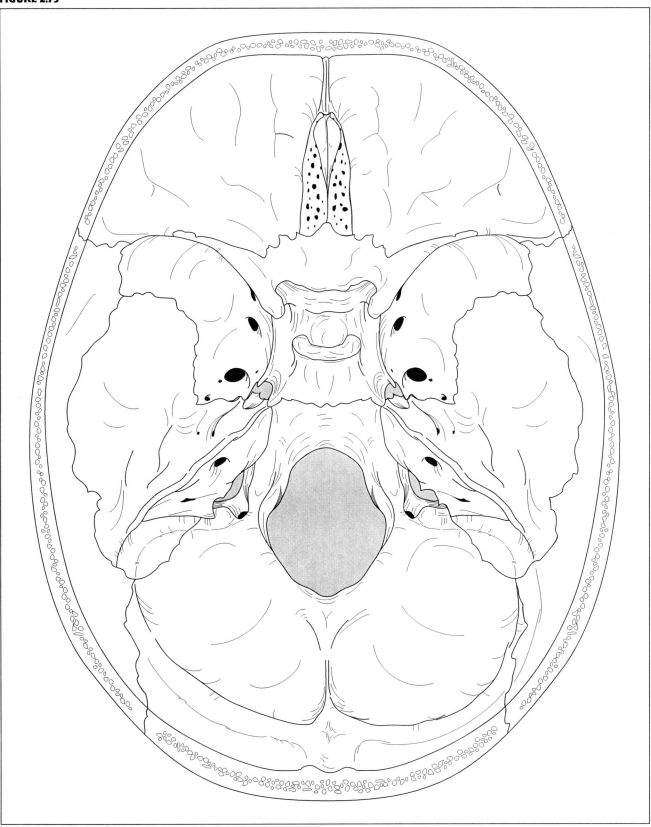

Illustration by Jamey Garbett. © 2003 Mark Nielsen.

FIGURE 2.76

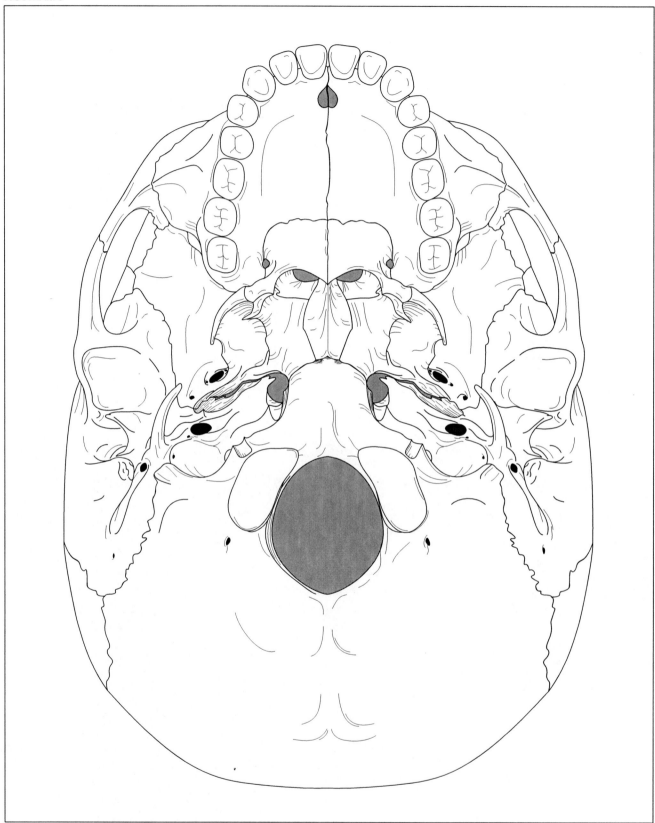

Axial Musculature

The axial skeleton musculature is composed of two major divisions. First, there are those muscles of the head and neck, which are collectively innervated by the cranial nerves. The remaining axial muscles include those of the trunk: the neck, thorax, and abdomen. These trunk muscles are innervated by spinal nerves rather than cranial nerves.

OBJECTIVES

- ☑ Identify the axial muscles
- ☑ Know the function of trunk muscles by group
- ☑ Know the functions particular to the individual muscles of the head and neck
- ☑ Understand the anatomical organization of the trunk's major muscle groups

STRUCTURES YOU ARE RESPONSIBLE FOR IDENTIFYING

Epaxial Muscles of the Trunk
Transversospinalis group
Erector spinae group

Hypaxial Muscles of the Trunk
Abdomen
Psoas major
Rectus abdominis
External oblique
Internal oblique
Transversus abdominis

Thorax
Longus colli
Serratus posterior muscles
Serratus anterior
Intercostal muscles (external, internal, innermost; transversus thoracis)
Diaphragm

Neck
Longus colli
Longus capitus
Strap muscles
Levator scapulae
Scalenes (anterior, middle, posterior)

Head/Neck Muscles
Extrinsic Eye Muscles
Superior rectus
Inferior rectus
Medial rectus
Lateral rectus
Superior oblique
Inferior oblique

Muscles of Mastication
Masseter
Temporalis
Medial and lateral pterygoids

Muscles of Facial Expression
Orbicularis oculi
Orbicularis ori
Frontalis
Occipitalis
Zygomaticus
Buccinator
Platysma

Neck Muscles (Cranial Nerve Innervated)
Sternocleidomastoid
Trapezius

Part 1. The Pattern of the Trunk

The illustration below shows a section through the abdomen, demonstrating the various muscle groups. We will see that the same pattern exists in the thorax and neck. The pattern even extends to the pelvis, but with much modification. The trunk musculature is broadly divided into epaxial and hypaxial muscles. The epaxial muscles are located posterior to the ribs and vertebrae.

The hypaxial muscles include all other trunk muscles. The first group to consider lies opposite from the epaxial muscles, along the ventral (anterior) side of the vertebrae: the subvertebral group. In the anterior wall of the trunk lies the ventral trunk musculature. The right and left lateral walls contain the lateral trunk musculature, which has four layers: the internal layer, middle layer, external layer, and outermost layer.

Identify the muscle groups in the figure below:

Expaxial muscles
Subvertebral
Ventral
Lateral (and its four layers)

FIGURE 3.1 Generalized pattern of the trunk

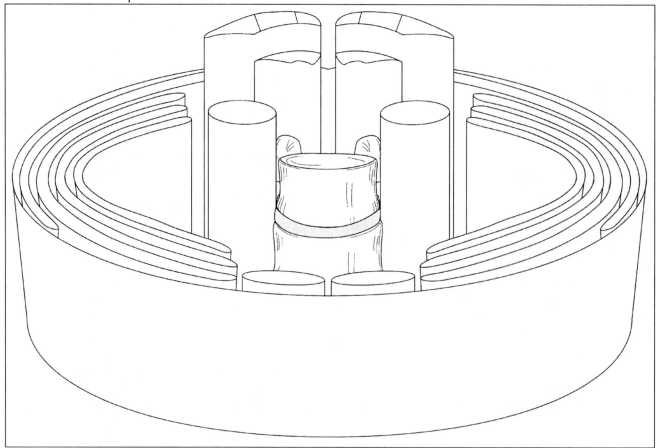

Illustration by Jamey Garbett. © 2003 Mark Nielsen.

Epaxial Muscles

The transversospinalis muscles lie in the groove created by the spinous process and the transverse process. They typically span two vertebrae and are responsible for rotation of the spine. The erector spinae muscles lie superficial to the transversospinalis muscles, and their fascicles run along the length of the spine and allow you to extend your spine, as when arching your back. The erector spinae and transversospinalis span the trunk from the sacrum to the neck.

FIGURE 3.2 Transversospinalis muscles

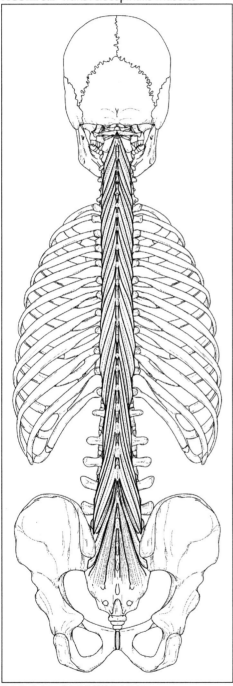

Illustration by Jamey Garbett. © 2003 Mark Nielsen.

FIGURE 3.3 Erector spinae muscles

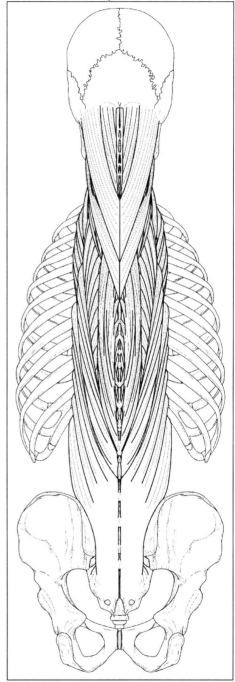

Illustration by Jamey Garbett. © 2003 Mark Nielsen.

Abdomen

Subvertebral Musculature

Psoas major

The subvertebral muscles of the abdomen are in a great position to flex the spine. In fact, many animals use these muscles for that very function. In about 60% of humans there is a small psoas minor that is a trunk flexor. However, in humans the major subvertebral muscle, the psoas major, passes out of the trunk and attaches to the femur. The psoas major and the iliacus from the pelvis combine to form a strong hip flexor, the iliopsoas muscle.

FIGURE 3.4

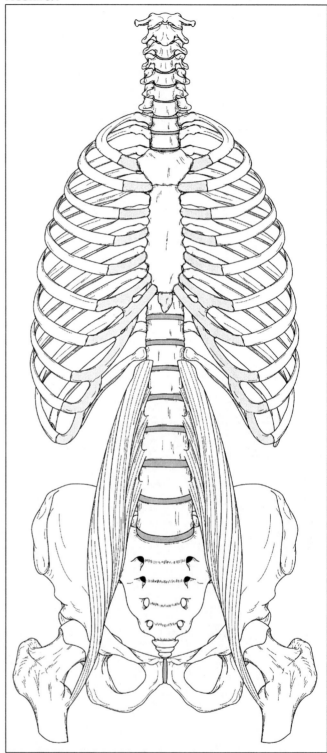

Illustration by Jamey Garbett. © 2003 Mark Nielsen.

Ventral Musculature

Rectus abdominis

The ventral muscles of the abdomen are flexors of the spine. The rectus abdominis allows you to do sit-ups and is often called the six-pack, in part because of its discrete, block-like segmentation. The rectus abdominis is surrounded by a thick connective tissue known as the rectus sheath that serves as an attachment for the lateral musculature.

FIGURE 3.5

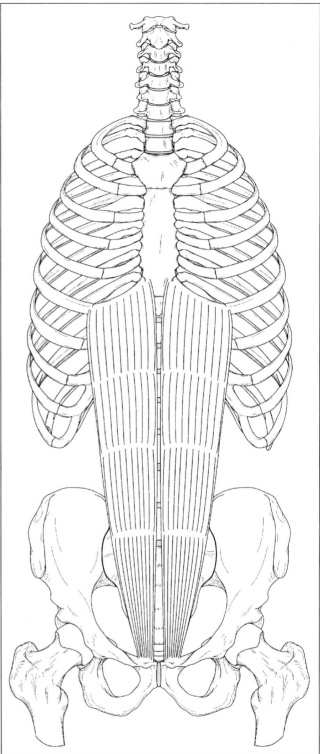

Illustration by Jamey Garbett. © 2003 Mark Nielsen.

Lateral Musculature: Internal Muscle Layer

Transversus abdominis

The lateral muscles of the abdomen rotate and flex the spine. Additionally, they increase intra-abdominal pressure, such as when occurs during coughing, defecation, urination, labor, or forced exhalation. The lateral musculature has fascicles oriented in different directions making a variety of movements possible, as well as making identification easier. The transversus abdominis muscle is the internal layer of the lateral wall and its fascicles run in a transverse plain.

FIGURE 3.6

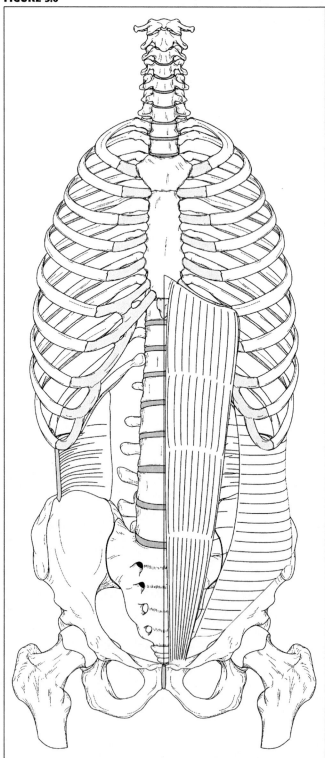

Illustration by Jamey Garbett. © 2003 Mark Nielsen.

Lateral Musculature: Middle Muscle Layer

Internal oblique

The internal oblique muscle is the middle layer of the lateral wall and its fascicles run inferiorly and laterally.

FIGURE 3.7

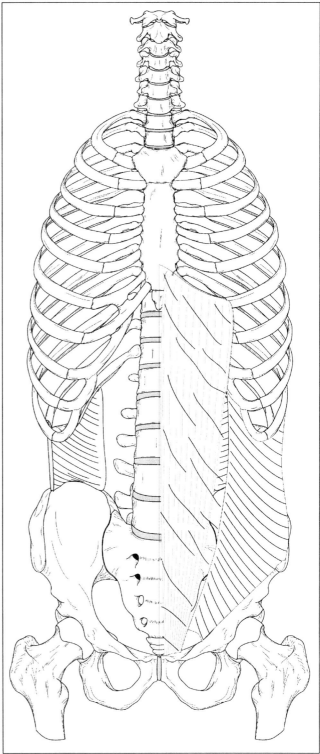

Illustration by Jamey Garbett. © 2003 Mark Nielsen.

Lateral Musculature: External and Outermost Muscle Layers

External oblique

The external oblique muscles are really two layers: one comprising the external and the other the outermost muscle layers of the four-layered trunk pattern. The fascicles run perpendicular to the internal oblique, superiorly and laterally.

FIGURE 3.8

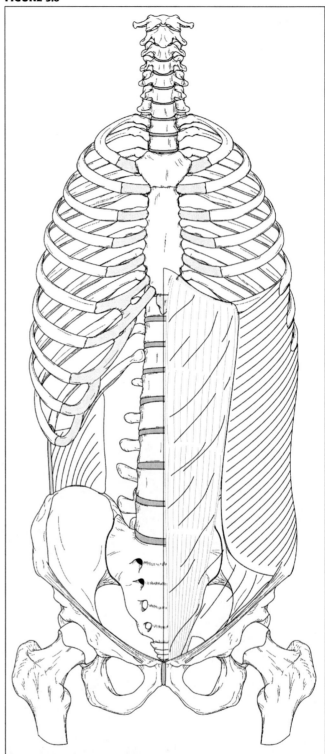

Illustration by Jamey Garbett. © 2003 Mark Nielsen.

Abdomen Summary

Using your knowledge of the abdominal wall, label the hypaxial muscles illustrated below: psoas major, rectus abdominis, transversus abdominis, internal oblique, external oblique (two layers).

FIGURE 3.9

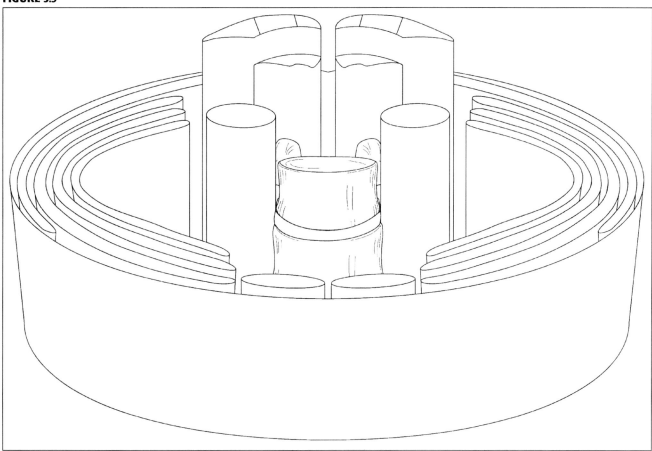

Thorax

Subvertebral Musculature

Longus colli

The subvertebral muscles of the thorax are present only in the upper thorax as the longus colli (colli = neck). Here, these muscles flex the neck. The subvertebral musculature is absent from the lower thorax where the rib cage prevents flexion of the spine.

FIGURE 3.10

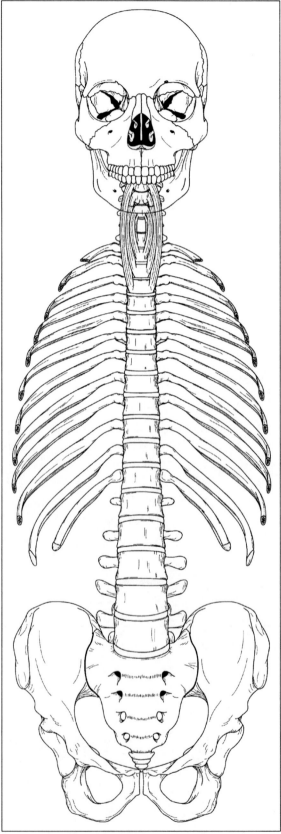

Illustration by Jamey Garbett. © 2003 Mark Nielsen.

Ventral Musculature

Sternalis muscle

The ventral muscles of the thorax are usually not present. When present (10%), the sternalis muscle runs from the sternum to the lower margin of the ribs and is not particularly useful for any movement.

Lateral Musculature: Internal, Middle, and External Muscle Layers

The intercostal muscles: innermost, internal, and external intercostals

The intercostal muscles span the spaces between adjacent ribs. They participate in respiratory movements, but more importantly they tense to provide a rigid wall to the thorax. Without this rigidity, the body wall would be sucked in and pushed out from between the ribs and diminish the efficiency of ventilation.

FIGURE 3.11
Intermost intercostal muscles

FIGURE 3.12
Internal intercostal muscles

FIGURE 3.13
External intercostal muscles

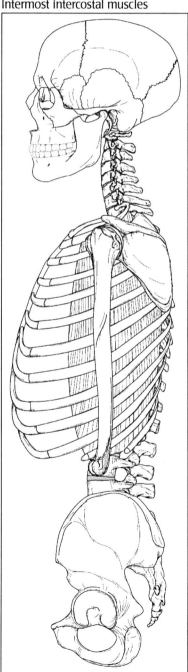
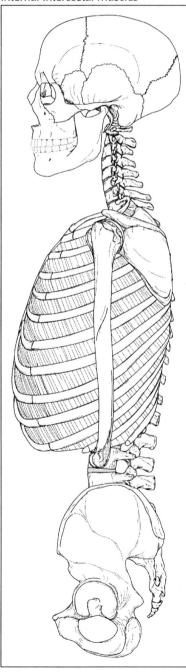
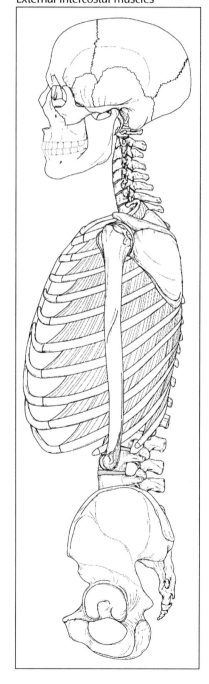

Lateral Musculature: Internal, Middle, and External Muscle Layers

The diaphragm

FIGURE 3.14 The diaphragm, part of the internal layer of the lateral musculature, at rest (left) and while contracting (right). Notice how these movements change the size of the thoracic cavity.

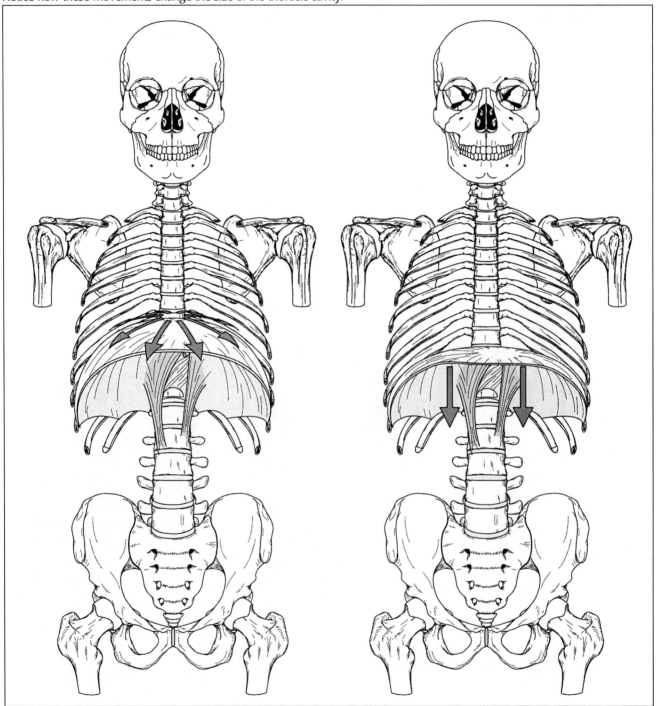

Illustration by Jamey Garbett. © 2003 Mark Nielsen.

Lateral Musculature: Outermost Layer

Serratus posterior muscles
Serratus anterior

The serratus posterior muscles (serratus posterior superior and serratus posterior inferior), unlike the intercostals, span several ribs and participate in moving the ribs during inspiration and expiration.

The serratus anterior muscle is a trunk muscle, but in the human it attaches to the scapula. It wraps around to connect the upper limb to the ribs. This muscle is often called the boxer's muscle; it pulls the scapula (and shoulder) forward and in doing so, extends the reach of the arm.

FIGURE 3.15 The serratus posterior muscles: serratus posterior superior and serratus posterior inferior

FIGURE 3.16 The serratus anterior muscle

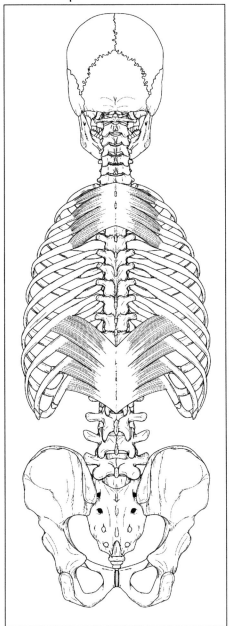

Illustration by Jamey Garbett. © 2003 Mark Nielsen.

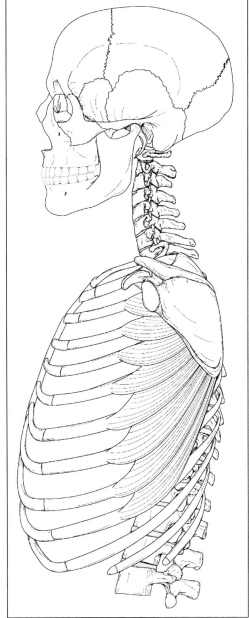

Illustration by Jamey Garbett. © 2003 Mark Nielsen.

Thorax Summary

Using your knowledge of the thoracic wall, label the hypaxial muscles illustrated in cross section below: longus colli, sternalis, intercostals (innermost, internal, external), serratus posterior, serratus anterior.

FIGURE 3.17

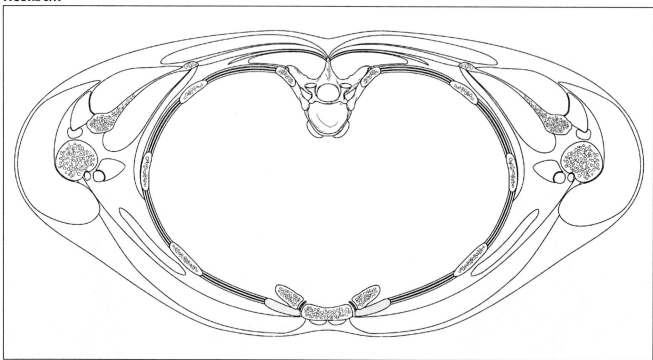

Illustration by Jamey Garbett. © 2003 Mark Nielsen.

Neck

Epaxial Musculature

The erector spinae and transversospinalis groups continue up into the neck. Superficially, there are additional muscles known as the splenius muscles that extend and rotate the neck.

Subvertebral Musculature

Longus colli and longus capitus

The longus colli begins in the thorax and continues up into the neck. Superiorly, the longus muscle attaches to the skull and is called longus capitus (colli = neck; capitus = head). These subvertebral muscles flex the neck and head.

FIGURE 3.18

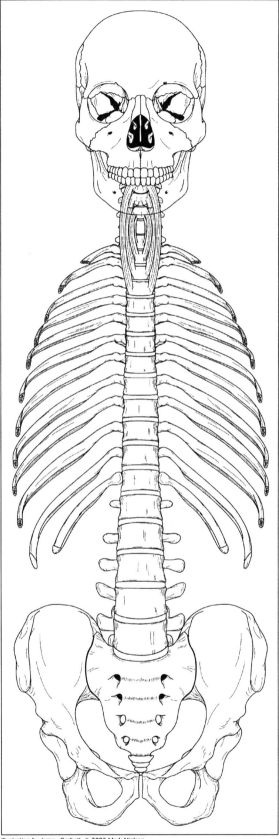

Illustration by Jamey Garbett. © 2003 Mark Nielsen.

Ventral Musculature

Strap muscles

Although the neck is not as obviously part of the trunk as are the abdomen and thorax, and it lacks a body cavity, the pattern of the trunk is evident. The organs of the neck separate the subvertebral musculature from the ventral musculature. Accordingly, the strap (infrahyoid) muscles lie ventral to the esophagus, hyoid, larynx, trachea, and thyroid gland. Generally, these muscles are responsible for movements of the hyoid and larynx that occur during swallowing.

FIGURE 3.19

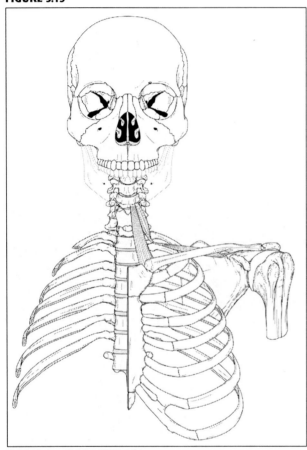

Illustration by Jamey Garbett. © 2003 Mark Nielsen.

FIGURE 3.20

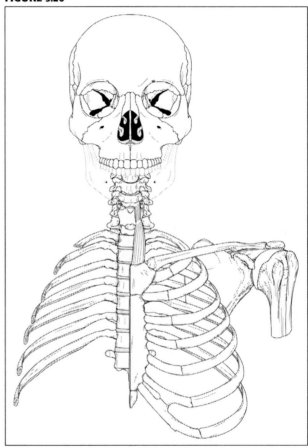

Illustration by Jamey Garbett. © 2003 Mark Nielsen.

Lateral Musculature: Internal, Middle, and External Muscle Layers

The scalene muscles: anterior, middle, and posterior scalenes

Lateral Musculature: Outermost Layer

Levator scapulae

The three scalene muscles attach from the transverse processes of cervical vertebrae to the first and second ribs and laterally flex the neck. When both sides contract simultaneously, they participate in forced inspiration by elevating the ribs. The outermost layer is named the levator scapulae. It too flexes the neck laterally. In addition, it elevates the scapula and shoulder as when shrugging.

FIGURE 3.21 Middle scalene muscle

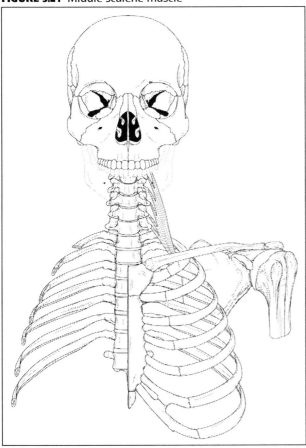

Illustration by Jamey Garbett. © 2003 Mark Nielsen.

FIGURE 3.22 Levator scapulae muscle

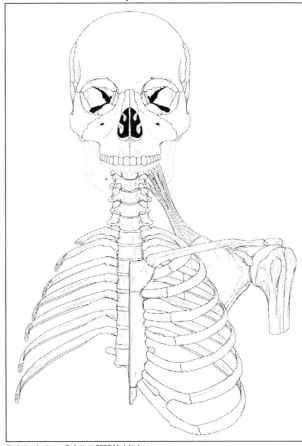

Illustration by Jamey Garbett. © 2003 Mark Nielsen.

Part 2. The Muscles of the Head and Neck

Embryologically, the origins of the muscles of the head are distinct from those of the trunk. However, in the adult, the line between these two regions becomes blurred at the neck. The pattern of innervation reveals the embryonic origins of muscles that have migrated out of their primitive position; trunk muscles are innervated by spinal nerves and head and neck muscles are innervated by cranial nerves.

Cranial nerve innervations of head and neck muscles:

Muscle group	Cranial nerve(s)
Extrinsic eye muscles	III, IV, and VI (oculomotor, trochlear, abducens)
Muscles of mastication	V (trigeminal)
Muscles of facial expression	VII (facial)
Pharynx and larynx*	IX, X (glossopharyngeal vagus)
Neck muscles	XI (accessory)
Tongue*	XII (hypoglossal)

*These muscles will not be examined

Extrinsic Eye Muscles

The four rectus muscles (superior rectus, inferior rectus, medial rectus, and lateral rectus) move the eyes in the vertical and horizontal axes. They emerge from the back of the orbit from a common tendinous ring and attach to the sclera (white) of the eye. The superior oblique and inferior oblique provide a small degree of rotation to the eye.

FIGURE 3.23

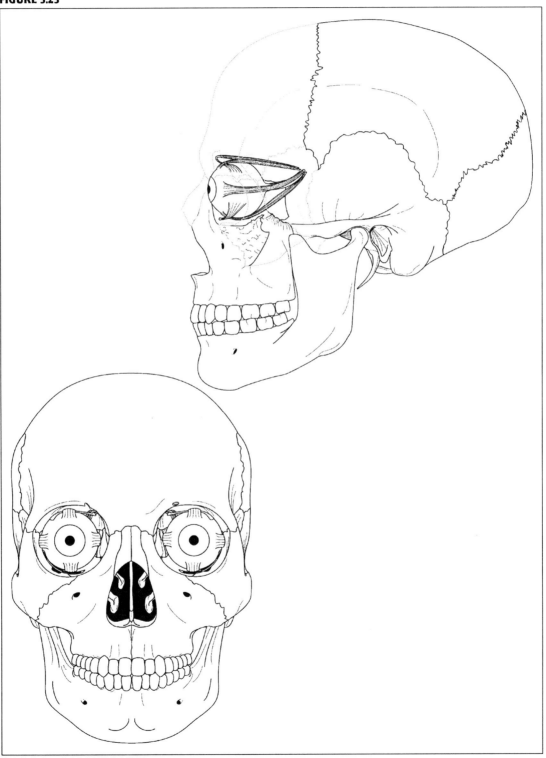

Muscles of Mastication

Temporalis
Masseter

These muscles are the prime closers (elevators) of the jaw. The temporalis attaches to the coronoid process of the mandible and to the frontal, parietal, and temporal bones of the skull. The masseter extends from the zygomatic arch down to the angle of the jaw.

FIGURE 3.24

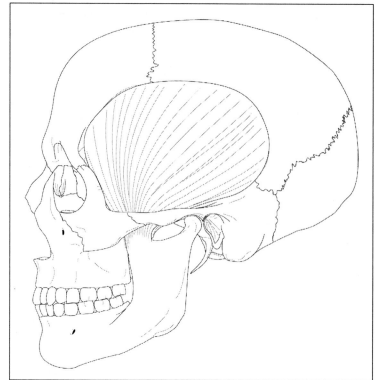

Illustration by Jamey Garbett. © 2003 Mark Nielsen.

FIGURE 3.25

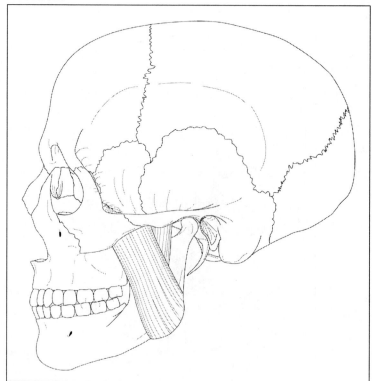

Illustration by Jamey Garbett. © 2003 Mark Nielsen.

Muscles of Mastication

Pterygoid muscles (medial pterygoid and lateral pterygoid)

These muscles are named for the pterygoid process of the sphenoid, to which they attach. Together, they produce side-to-side movements of the mandible when chewing. These are difficult to see on most models since they reside deep to the zygomatic arch, masseter, and temporalis.

FIGURE 3.26 Medial pterygoid muscle

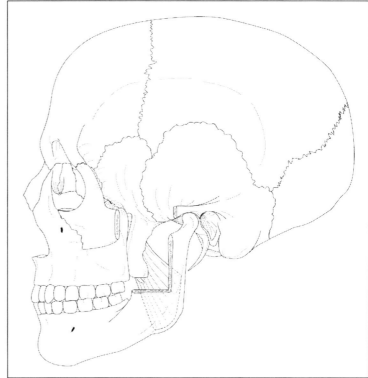

Illustration by Jamey Garbett. © 2003 Mark Nielsen.

FIGURE 3.27 Lateral pterygoid muscle

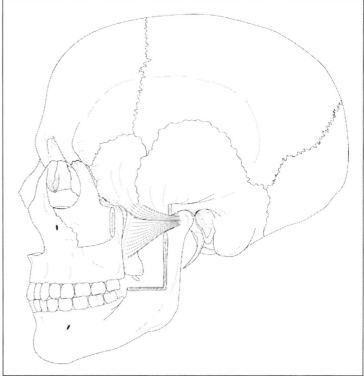

Illustration by Jamey Garbett. © 2003 Mark Nielsen.

Muscles of Facial Expression

Orbicularis oculi
Orbicularis ori
Occipitofrontalis (occipitalis frontalis)
Zygomaticus
Buccinator
Platysma

These muscles are superficial muscles that move the face. They are, in some cases, attached only to the skin of the face. There are several more very small muscles of facial expression not listed, including those that move the ear and wrinkle the skin of the nose and other small movements of the skin. The platysma is a large thin muscle that is attached to bone, being responsible for depressing (opening) the jaw. It covers the entire anterior surface of the neck. The names of the muscles give a clear indication of their location. What does Buccinator refer to?

FIGURE 3.28

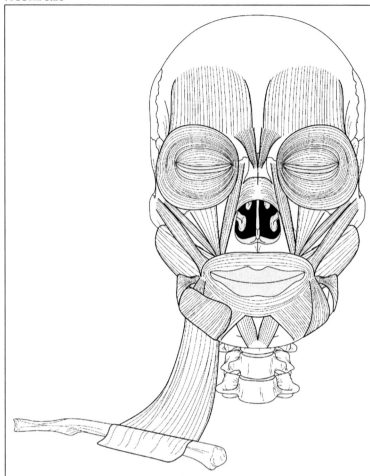

Illustration by Jamey Garbett. © 2003 Mark Nielsen.

Neck Muscles
(Cranial-Nerve-Innervated)

Trapezius

The cranial-nerve-innervated neck muscles originate near the other cranial muscles but over the course of development come to lie in more caudal locations. The trapezius, for example, takes attachment from the lower thoracic vertebrae. This muscle participates in head and neck movements, but the primary role of the trapezius is the support of the upper limb. When you are carrying a suitcase, for example, your trapezius attempts to maintain the elevation of your shoulder.

FIGURE 3.29

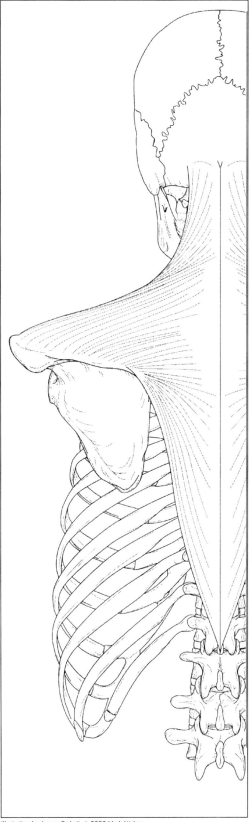

Illustration by Jamey Garbett. © 2003 Mark Nielsen.

Neck Muscles
(Cranial-Nerve-Innervated)

Sternocleidomastoid

The sternocleidomastoid is named for its three major attachment sites. This muscle causes rotation of the head to the contralateral side as the mastoid process is pulled toward the sternum. You can easily see this muscle as a person turns their head to the side. Both muscles acting together flex the neck and head.

FIGURE 3.30

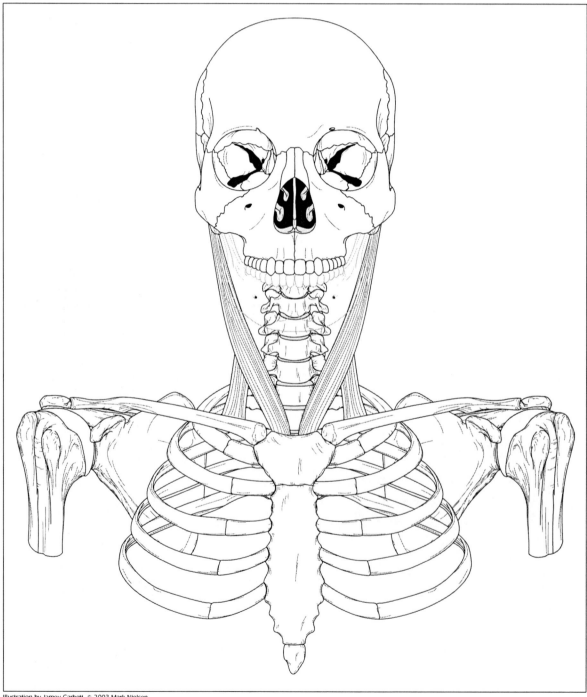

Illustration by Jamey Garbett. © 2003 Mark Nielsen.

Axial Musculature

Use the illustrations below to test your understanding of muscles.

FIGURE 3.31

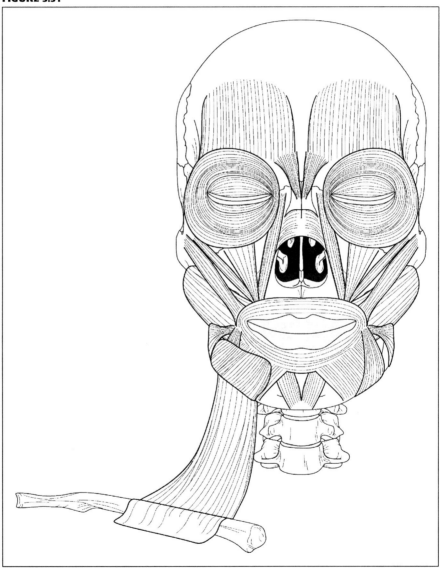

Illustration by Jamey Garbett. © 2003 Mark Nielsen.

Identify the longus coli, the anterior, middle, and posterior scalenes, and the strap muscles. Which groups do these muscles belong to?

FIGURE 3.32

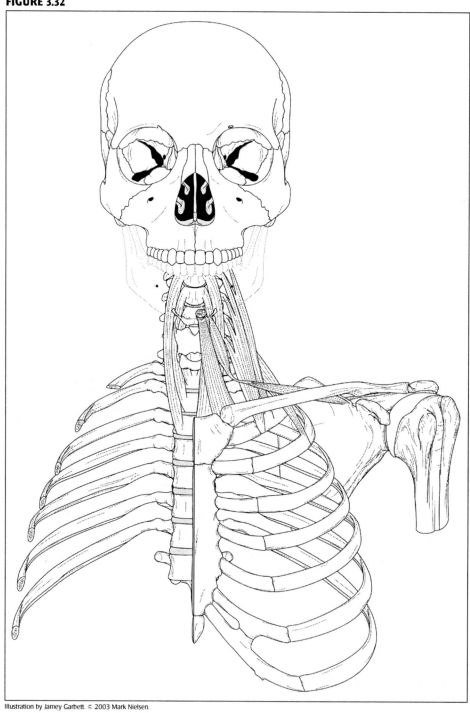

On the following diagram, label each group of muscles. Then, under each of these group headings, list the names of the muscles for the abdomen, thorax, and neck.

FIGURE 3.33

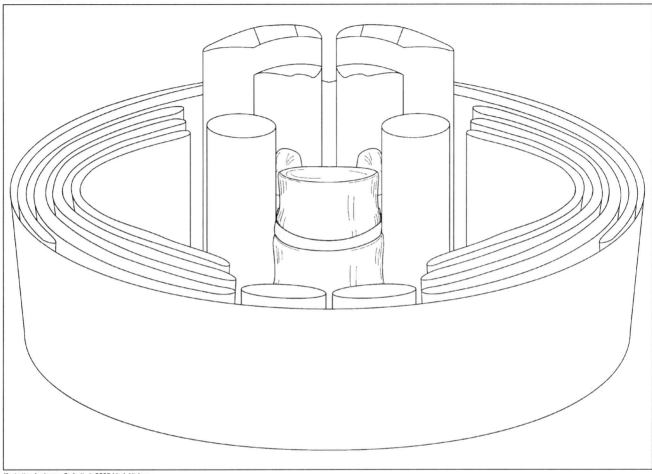

Illustration by Jamey Garbett. © 2003 Mark Nielsen.

FIGURE 3.35

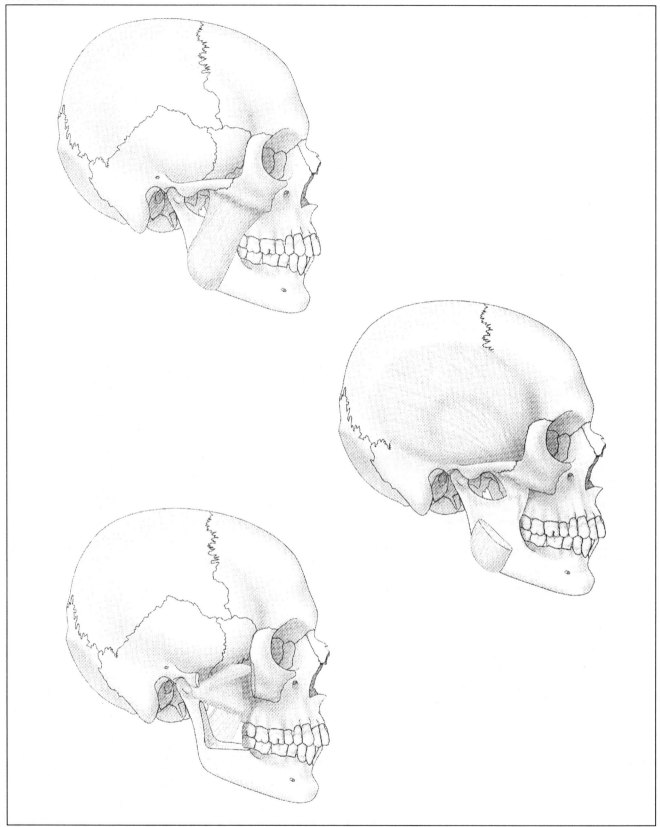

From *Laboratory Guide for Human Anatomy* by William J. Radke, copyright © 2002 John Wiley & Sons, Inc. Reprinted by permission of John Wiley & Sons, Inc.

Upper Limb

The upper limb bones include the phalanges, the metacarpals, the carpals, the radius, the ulna, and the humerus. Additionally, the two bones of the pectoral girdle, the scapula and clavicle, are considered part of the upper limb. The pectoral girdle connects the upper limb to the axial skeleton at the sternoclavicular joint. Despite this single connection, the scapula is held securely by several muscles, as we will soon see.

OBJECTIVES

☑ Identify the bones of the upper limb and the names of the joints where they articulate
☑ Know the various movements that can occur at each joint
☑ Identify the muscles of the upper limb and know their function

STRUCTURES YOU ARE RESPONSIBLE FOR IDENTIFYING

Bones and Landmarks
Clavicle
 Sternal end
 Acromial end
Scapula
 Coracoid process
 Acromion
 Spine
 Supraspinous fossa
 Infraspinous fossa
 Subscapular fossa
 Glenoid cavity
Humerus
 Head
 Shaft
 Tubercles (greater and lesser)
 Intertubercular groove
 Deltoid tuberosity
 Medial epicondyle
 Lateral epicondyle
 Trochlea
 Capitulum
Radius
 Head
 Radial tuberosity
 Styloid process
Ulna
 Coronoid process
 Ulnar tuberosity

 Styloid process
 Olecranon process
Carpal Bones
 Scaphoid
 Lunate
 Triquetrium
 Pisiform
 Trapezium
 Trapezoid
 Capate
 Hamate
Metacarpal Bones
Phalanges
Joints
 Sternoclavicular joint
 Glenoid
 Humeroulnar joint
 Humeroradial joint
 Proximal radioulnar joint
 Distal radioulnar joint
Muscles
 Rhomboids (major and minor)
 Levator scapulae
 Trapezius
 Serratus anterior
 Pectoralis major
 Supraspinatus
 Infraspinatus

Teres minor
Subscapularis
Deltoid
Teres major
Latissimus dorsi
Coracobrachialis
Brachialis
Biceps brachii
Triceps brachii
Pronator teres
Flexor carpi radialis
Palmaris longus
Flexor carpi ulnaris
Flexor digitorum (superficialis
 and profundus)
Flexor pollicis longus
Pronator quadratus
Brachioradialis
Extensor carpi radialis (longus
 and brevis)
Extensor digitorum
Extensor digiti minimi
Extensor carpi ulnaris
Supinator

The Skeletal System

The skeleton can be divided into two basic divisions: the appendicular skeleton and the axial skeleton. The appendicular skeleton, as its name suggests, consists of the bones of the upper and lower limbs or appendages and the associated pectoral and pelvic girdles. The axial skeleton forms the central axis of the body and consists of the vertebral elements, ribs, sternum, and skull.

FIGURE 4.1

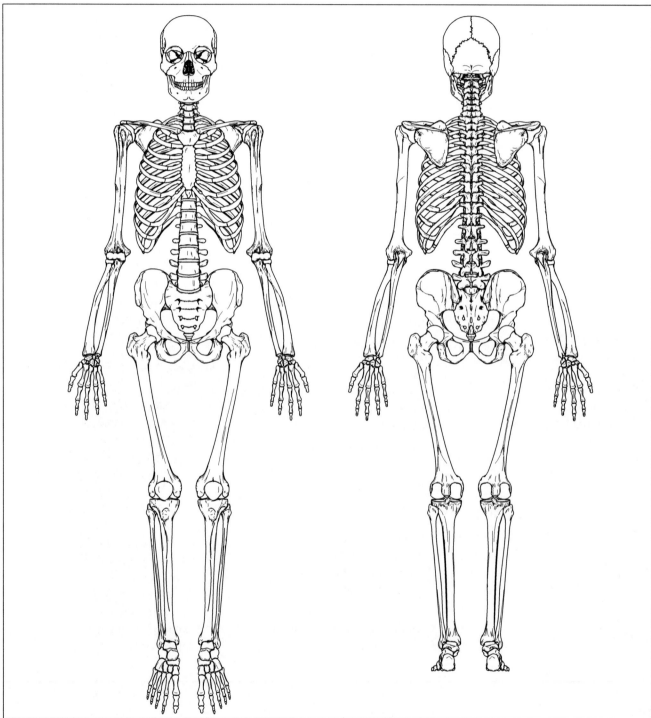

Illustration by Jamey Garbett. © 2003 Mark Nielsen.

From *Human Anatomy: Lab Manual and Workbook, 4th edition* by Mark Nielsen. Copyright 2002 by Mark Nielsen. Reprinted by permission of Kendall/Hunt Publishing Company.

Pectoral Girdle

The pectoral girdle is formed by two bones, the clavicle and the scapula. Neither of the bones have a direct articulation with the vertebral column and there is no direct articulation between right and left pectoral girdles. The pectoral girdle is comparatively light weight and forms a shallow joint with the limb. Compare this with the anatomy of the pelvic girdle. What is the functional significance for the differences between the two girdles?

Clavicle

clavis = little key

The anterior bone of the pectoral girdle.

FIGURE 4.2 Right clavicle, superior view (top), Right clavicle, inferior view (bottom)

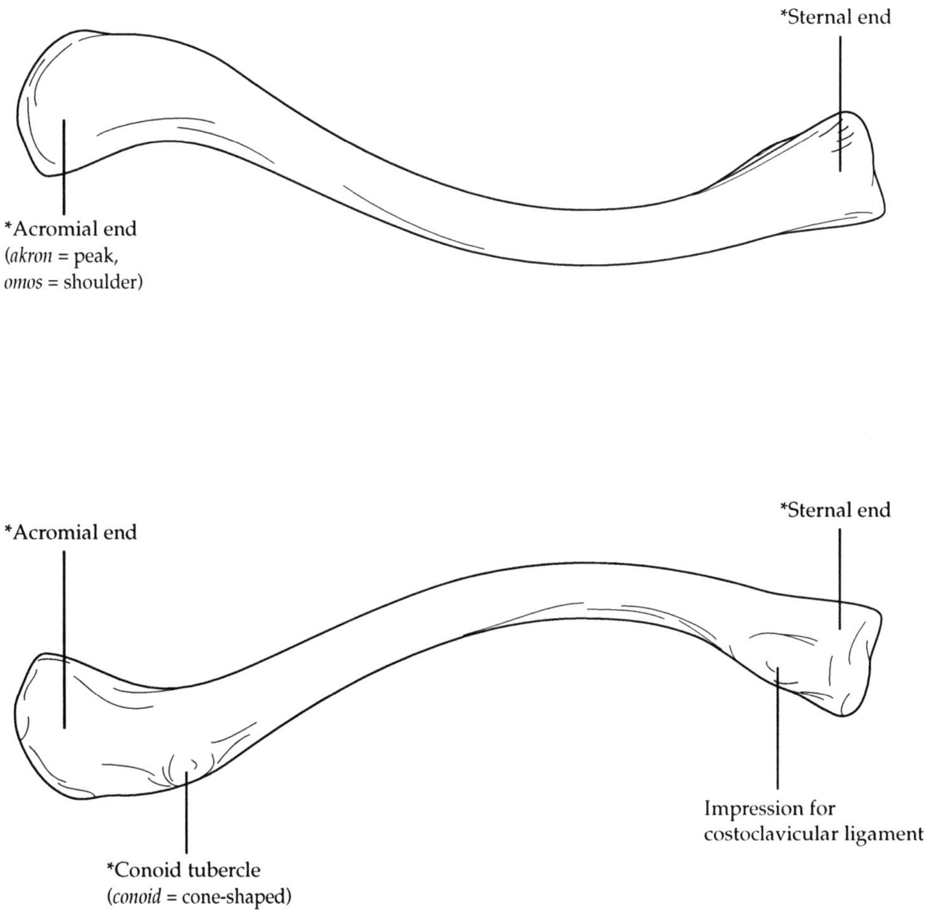

*Sternal end

*Acromial end
(*akron* = peak,
omos = shoulder)

*Sternal end

*Acromial end

Impression for
costoclavicular ligament

*Conoid tubercle
(*conoid* = cone-shaped)

*Indicates key landmarks to orient the bone. Familiarize yourself with these terms before the first lab.

Illustration by Jamey Garbett. © 2003 Mark Nielsen.

Scapula

scapter = spade or trowel

The posterior bone of the pectoral girdle.

FIGURE 4.3 Right scapula, posterior view

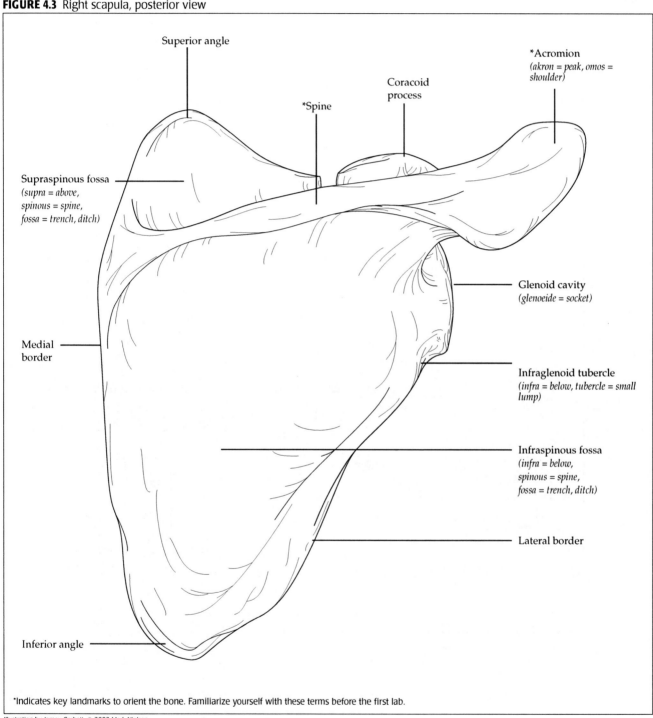

Superior angle

*Acromion
(akron = peak, omos = shoulder)

Coracoid process

*Spine

Supraspinous fossa
(supra = above, spinous = spine, fossa = trench, ditch)

Glenoid cavity
(glenoeide = socket)

Medial border

Infraglenoid tubercle
(infra = below, tubercle = small lump)

Infraspinous fossa
(infra = below, spinous = spine, fossa = trench, ditch)

Lateral border

Inferior angle

*Indicates key landmarks to orient the bone. Familiarize yourself with these terms before the first lab.

Illustration by Jamey Garbett. © 2003 Mark Nielsen.

Scapula

FIGURE 4.4 Right scapula, anterior view

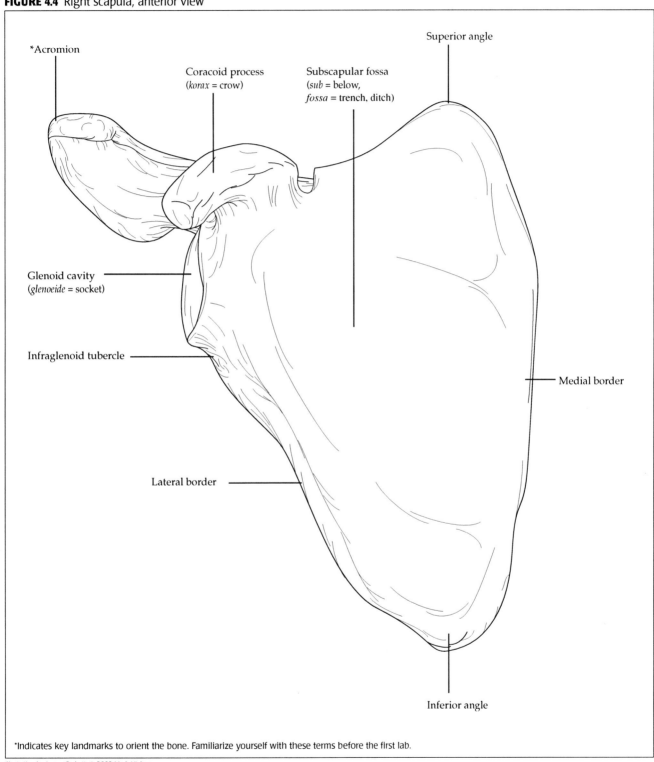

*Acromion

Coracoid process
(*korax* = crow)

Subscapular fossa
(*sub* = below,
fossa = trench, ditch)

Superior angle

Glenoid cavity
(*glenoeide* = socket)

Infraglenoid tubercle

Medial border

Lateral border

Inferior angle

*Indicates key landmarks to orient the bone. Familiarize yourself with these terms before the first lab.

Illustration by Jamey Garbett. © 2003 Mark Nielsen.

The Upper Limb Proper

The superior limb proper consists of three basic regions—brachium (arm), antebrachium (forearm), and manus (hand). There are a total of 32 bones in each upper limb (pectoral girdle and superior limb proper combined).

Humerus

related to omos = shoulder

The bone of the brachium.

FIGURE 4.5 Right humerus, anterior view (left), Right humerus, posterior view (right)

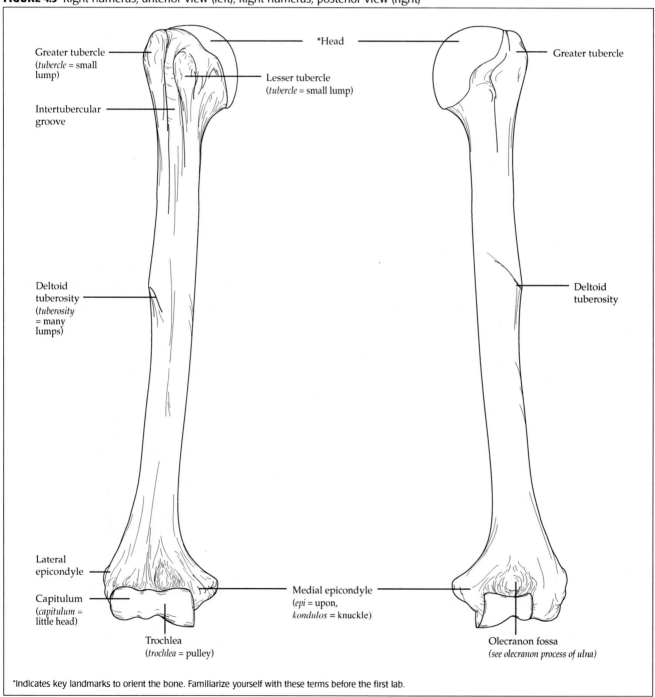

Greater tubercle
(*tubercle* = small lump)

Intertubercular groove

Deltoid tuberosity
(*tuberosity* = many lumps)

Lateral epicondyle

Capitulum
(*capitulum* = little head)

*Head

Lesser tubercle
(*tubercle* = small lump)

Trochlea
(*trochlea* = pulley)

Medial epicondyle
(*epi* = upon, *kondulos* = knuckle)

Greater tubercle

Deltoid tuberosity

Olecranon fossa
(*see olecranon process of ulna*)

*Indicates key landmarks to orient the bone. Familiarize yourself with these terms before the first lab.

Illustration by Jamey Garbett. © 2003 Mark Nielsen.

Ulna

olena = forearm

The medial bone of the antebrachium.

Radius

radi = stake or spoke

The lateral bone of the antebrachium.

FIGURE 4.6 Anterior view (left), Posterior view (right)

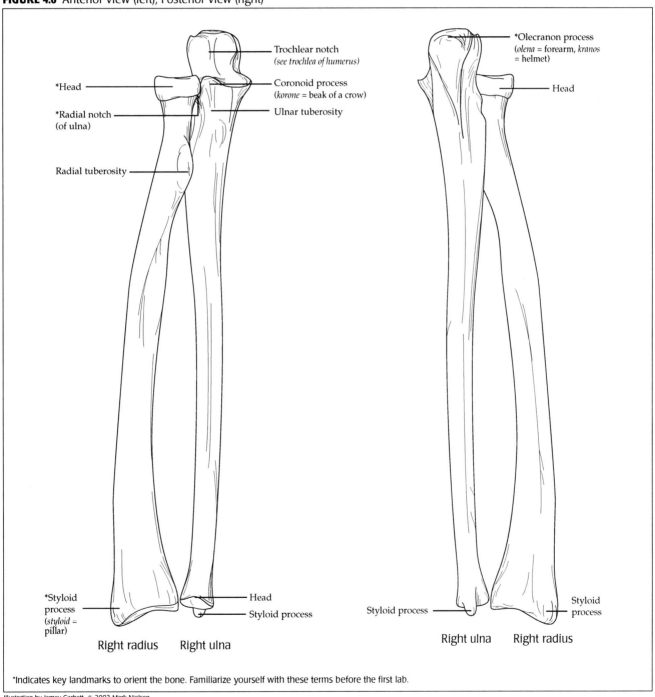

Trochlear notch
(see trochlea of humerus)

*Head

*Radial notch
(of ulna)

Radial tuberosity

Coronoid process
(korone = beak of a crow)

Ulnar tuberosity

*Olecranon process
(olena = forearm, kranos = helmet)

Head

*Styloid process
(styloid = pillar)

Head

Styloid process

Styloid process

Styloid process

Right radius   Right ulna

Right ulna   Right radius

*Indicates key landmarks to orient the bone. Familiarize yourself with these terms before the first lab.

Illustration by Jamey Garbett. © 2003 Mark Nielsen.

Hand or Manus

The hand skeleton consists of three regions: the carpus or wrist, the metacarpal region or palm, and the digits or fingers and thumb.

Carpal Bones

karpos = wrist

The eight small short bones, called wrist bones, at the proximal end of the hand skeleton.

Metacarpal Bones

meta = beyond, next to; karpos = wrist

The five short long bones that form the palm region of the hand skeleton.

Phalanges

phalanx = soldiers lined up for battle

The fourteen bones of the digits (fingers and thumb). Each finger has three phalanges, while the thumb has only two.

FIGURE 4.7 Right hand, anterior view

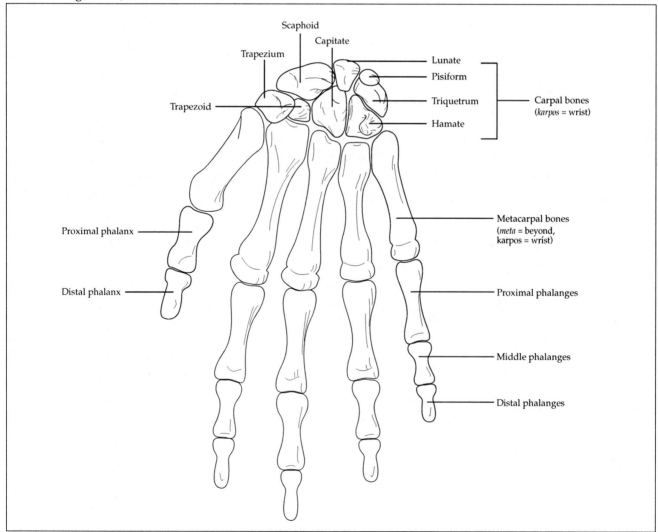

Illustration by Jamey Garbett. © 2003 Mark Nielsen.

Learning Muscles

The following ideas, if used, will help you learn the anatomy of the muscles. By following the steps outlined below, you will introduce more association and logic into the memorization process. Introducing these features into the memorization process will enhance your ability to recall the information and retain it in long term memory.

Picture the Muscle

FIGURE 4.8

Visualize the muscle. Study a picture of the individual muscle and visually note where it attaches to the skeleton. If you can form a strong mental image of the muscle, if you can visualize its relation to the skeleton, and picture its position on the body, then the learning process becomes a descriptive one and not one of sheer memorization of lists and tables of words. Look at these pictures of a muscle called the latissimus dorsi and describe, in your own words the relationships it has to the skeleton, the direction its fibers are running, how and which joints it is crossing.

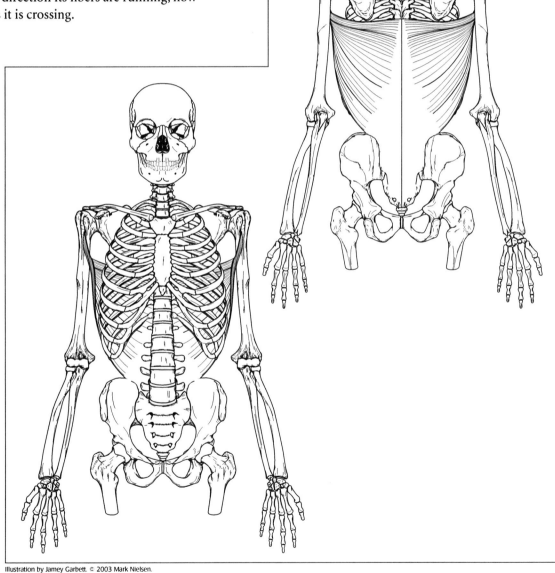

Illustration by Jamey Garbett. © 2003 Mark Nielsen.

From *Human Anatomy: Lecture Manual, 4th edition* by Mark Nielsen, illustrations by Jamey Garbett. Copyright © 2002 by Mark Nielsen. Reprinted by permission of Kendall/Hunt Publishing Company.

Make the Muscle Name Work for You

Keep in mind that the full name of any muscle of the body begins with musculus, a masculine noun. The English language has dropped this from common usage and allows the adjective and the genitive that describe the muscle to stand alone. Thus, musculus latissimus dorsi becomes latissimus dorsi. Muscles were originally named by describing their various characteristics: shape, size, location, or actions. So if you analyze musculus latissimus dorsi you see:

- ☐ musculus = muscle – noun being described;
- ☐ latissimus = broadest – adjective describing the noun;
- ☐ dorsi = of the back – genitive form of dorsum referring to the part of the body where the muscle is located;

Therefore, latissimus dorsi equates to the broadest muscle of the back, which you notice, from the picture you studied on the previous page, is an excellent description of the muscle.

So by using the etymological process, you can learn a great deal about the muscle from its name alone.

See and Describe the Muscle's Attachments

If you have a good visual image of the muscle, learning the muscle attachments becomes a descriptive process rather than one that involves sheer memorization of words. Simply describe your visual picture of the muscle attaching to the named parts of the skeleton. (This will require a good knowledge of the skeletal system. For this reason, you gained a good foundation of the skeletal system earlier in the course.) As you look at the latissimus dorsi you can see that it comes from the spinous processes of the vertebrae from mid-thorax down into the sacrum and attaches at the other end on the intertubercular groove of the humerus.

Learning the Movements Muscles Produce

With a good visual picture of the muscle, understanding the joint movements it produces should follow fairly naturally. Keep the following in mind when thinking about the movements muscles produce:

1. Muscles pull on a straight line bringing the sites of attachment closer together. (This is why it is important to understand muscle attachments.)
2. Have a clear picture of the path of the muscle as it crosses a joint, especially at the freely movable ball and socket joints.
3. A muscle produces an action at every joint it crosses.

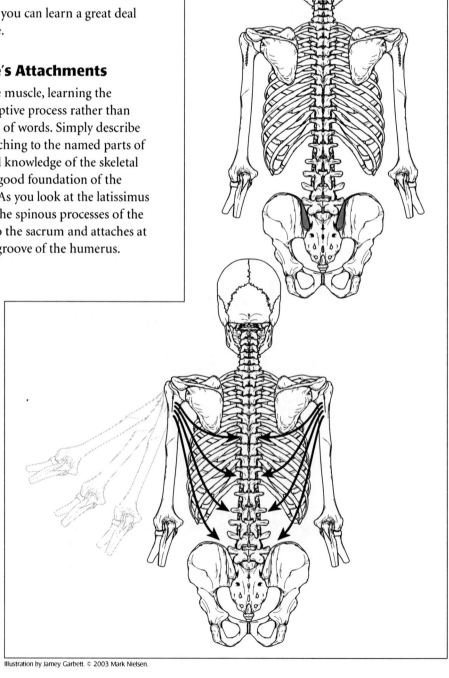

FIGURE 4.9

Illustration by Jamey Garbett. © 2003 Mark Nielsen.

Group Muscles That Share Attachments and Actions in Common

Many muscles have attachments and actions in common. Grouping muscles in this way encourages thought and simplifies learning. Much of this grouping is a natural phenomenon of the body's development.

For example, the latissimus dorsi muscle shares a common attachment with the teres major and pectoralis major on the intertubercular groove of the humerus and all the muscles are effective adductors of the shoulder joint.

FIGURE 4.10

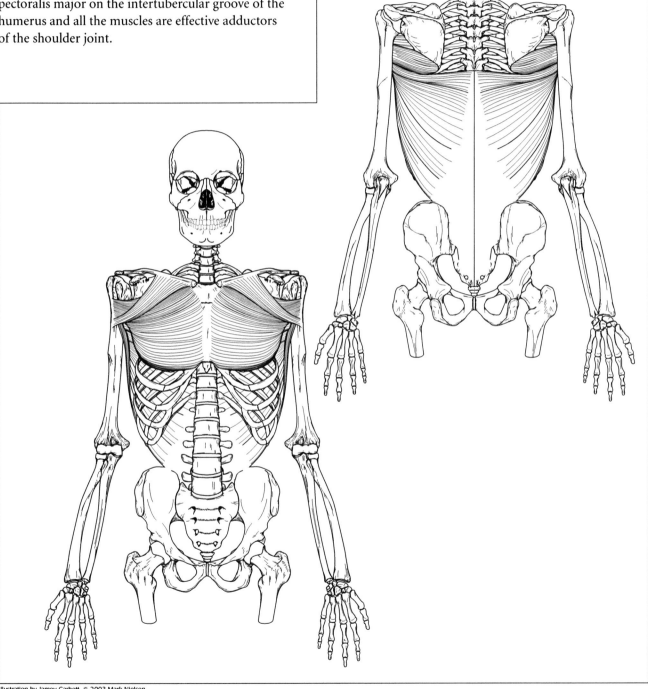

Illustration by Jamey Garbett. © 2003 Mark Nielsen.

Scapular Anatomy

Review of the Scapula

Attachments to the axial skeleton

FIGURE 4.11

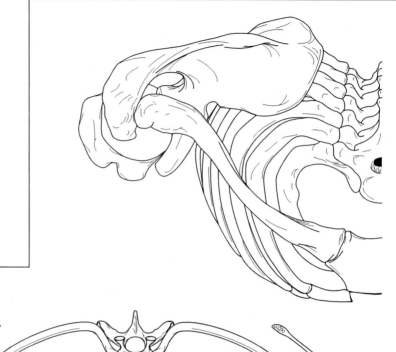

Muscular sling

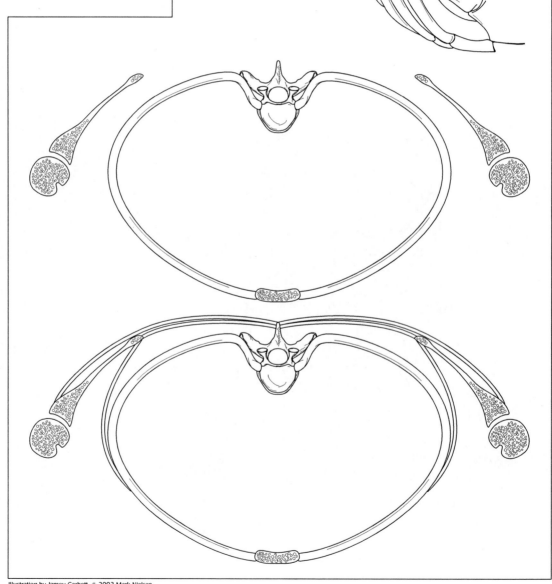

Illustration by Jamey Garbett. © 2003 Mark Nielsen.

Scapular Movements

FIGURE 4.12

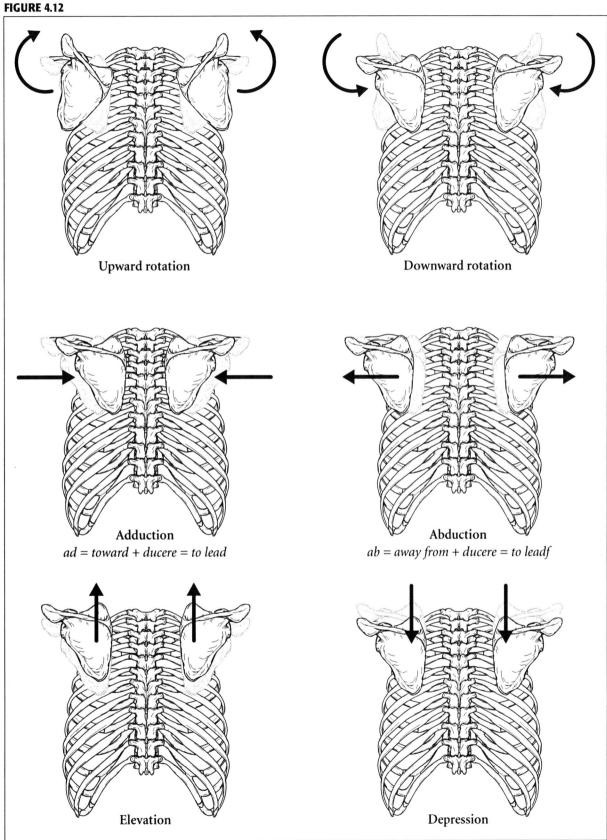

Upward rotation

Downward rotation

Adduction
ad = toward + ducere = to lead

Abduction
ab = away from + ducere = to leadf

Elevation

Depression

Scapular Muscles

Rhomboideus Major Muscle

rhomboid = a parallelogram with unequal adjacent sides, major = greater

Origin

Insertion

Function

Structure and relationships

FIGURE 4.13

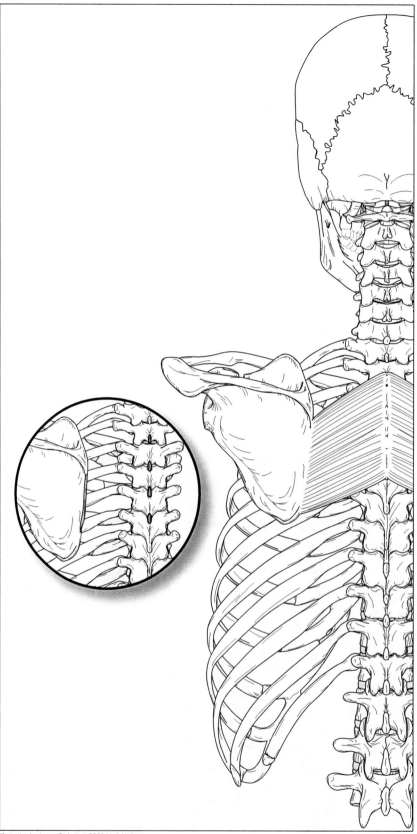

Illustration by Jamey Garbett. © 2003 Mark Nielsen.

Rhomboideus Minor Muscle

rhomboid = a parallelogram with unequal adjacent sides, **minor** *= lesser*

Origin

Insertion

Function

Structure and relationships

FIGURE 4.14

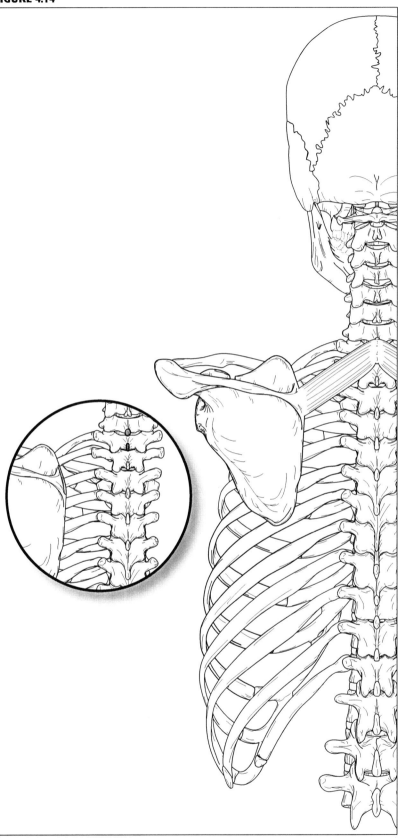

Illustration by Jamey Garbett. © 2003 Mark Nielsen.

Levator Scapulae Muscle

levator = one that lifts or elevates

Origin

Insertion

Function

Structure and relationships

FIGURE 4.15

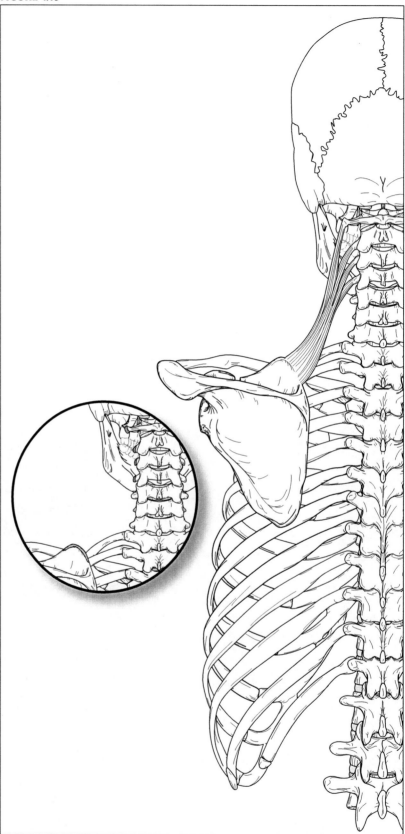

Illustration by Jamey Garbett. © 2003 Mark Nielsen.

Trapezius Muscle

trapezion = an irregular four sided figure

Origin

Insertion

Function

Structure and relationships

FIGURE 4.16

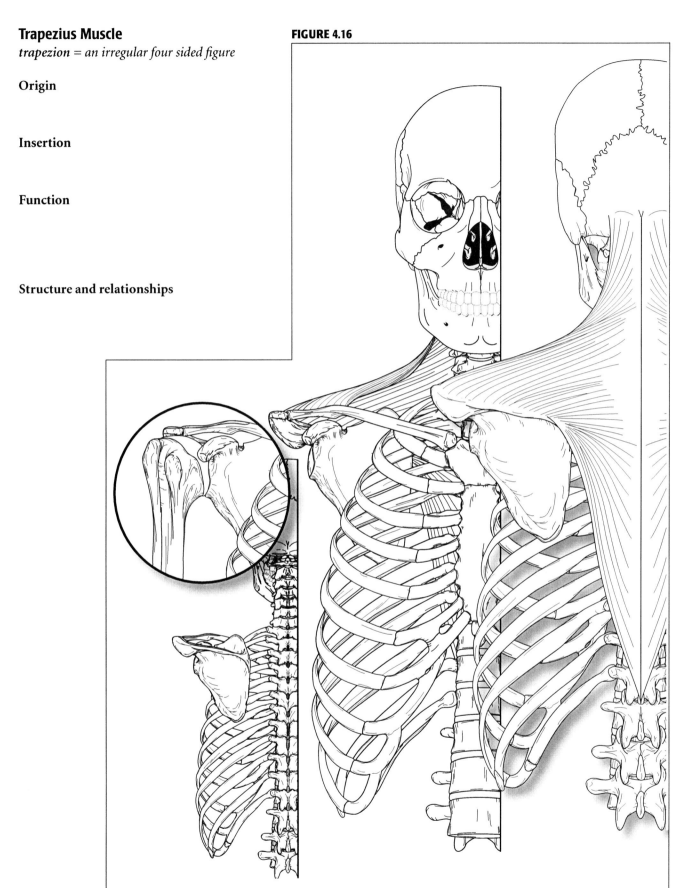

Illustration by Jamey Garbett. © 2003 Mark Nielsen.

Serratus Anterior Muscle

serra = saw

Origin

Insertion

Function

Structure and relationships

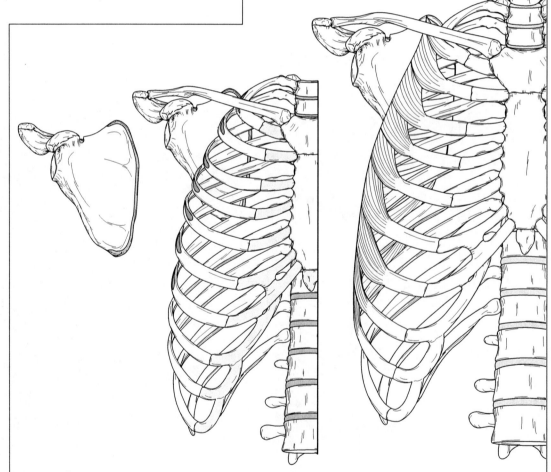

Illustration by Jamey Garbett. © 2003 Mark Nielsen.

FIGURE 4.17

Serratus Anterior Muscle

FIGURE 4.18

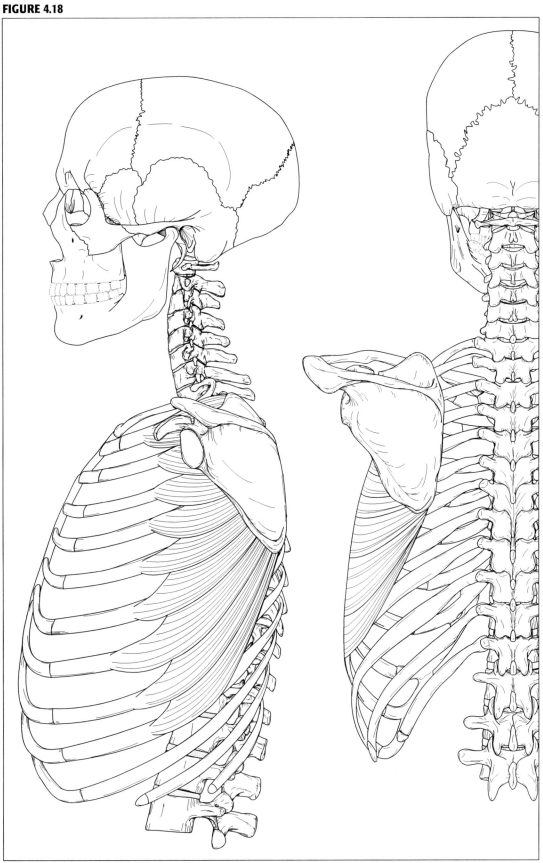

Illustration by Jamey Garbett. © 2003 Mark Nielsen.

Pectoralis Minor Muscle

pectus = the breast, minor = lesser

Origin

Insertion

Function

Structure and relationships

FIGURE 4.19

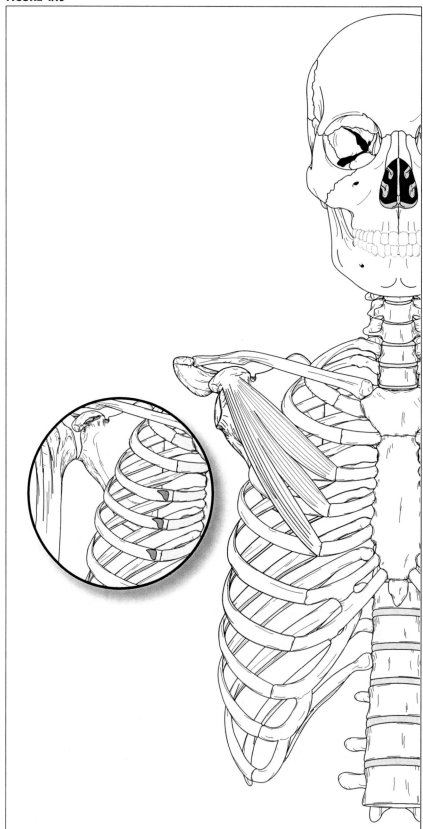

Illustration by Jamey Garbett. © 2003 Mark Nielsen.

Subclavius Muscle

sub = below, clavius for clavicle = little key

Origin

Insertion

Function

Structure and relationships

FIGURE 4.20

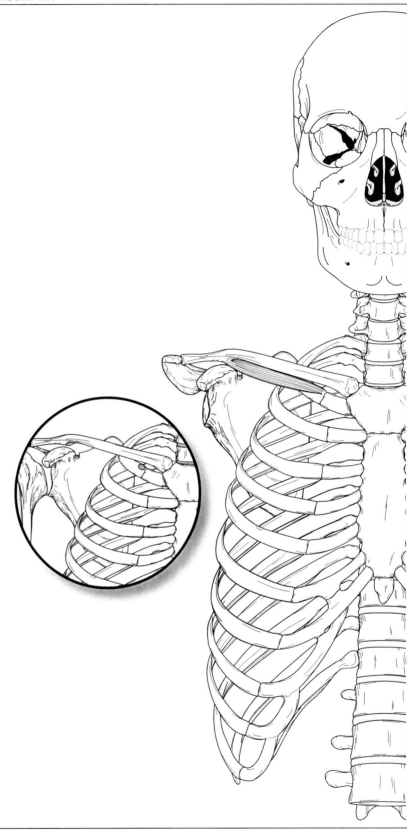

Illustration by Jamey Garbett. © 2003 Mark Nielsen.

Summary of the Scapular Sling Muscles

Common attachments

Common actions and function

Practical applications

FIGURE 4.21

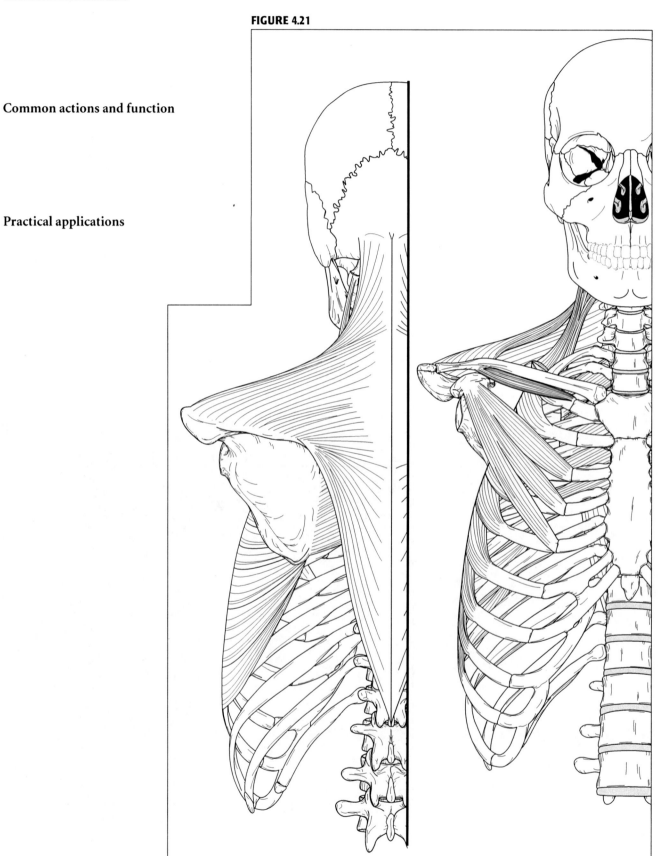

Illustration by Jamey Garbett. © 2003 Mark Nielsen.

Muscles of the Scapula

Muscle	Origin	Insertion	Function	Innervation
Rhomboideus major	Spinous processes of thoracic vertebrae 2 to 5	Medial margin of scapula between spine and inferior angle	Retraction (adduction) of scapula; downward rotation of scapula; stabilize the scapula	Dorsal scapular nerve (C4 and C5)
Rhomboideus minor	Lower end of nuchal ligament and spinous processes of C7 to T1	Medial margin of scapula at the base of the scapular spine and towards the superior angle	Retraction (adduction) of scapula; downward rotation of scapula; stabilize the scapula	Dorsal scapular nerve (C4 and C5)
Levator scapulae	Posterior tubercles of the transverse processes of cervical vertebrae 1 to 4	Superior angle and upper medial margin of the scapula	Elevation of scapula; lateral flexion of the neck; stabilize the scapula	Branches from C3 and C4 ventral rami and from C5 via dorsal scapular nerve
Trapezius	Superior nuchal line near midline, external occipital protuberance, nuchal ligament, spinous processes of 7th cervical and all thoracic vertebrae	Posterior border of lateral third of clavicle, medial border of acromion superior edge and medial base of scapular spine	Upward rotation of scapula; assists with elevation; retraction (adduction) of scapula; extension of head; stabilize the scapula	Accessory nerve (Cranial nerve XI)
Serratus anterior	Outer surface of first 8 or 9 ribs just anterior to mid-axillary line and the intercostal fascia	Medial margin of scapula from superior angle to inferior angle, at the superior and inferior angles the fibers pass onto the dorsal surfaces of the respective angles	Protraction (abduction) of scapula; upward rotation of scapula; Upper fibers help suspend the scapula; stabilize the scapula	Long thoracic nerve (C5, C6, and C7)
Pectoralis minor	Junction of ribs 3, 4, and 5 with their cartilage and fascia of external intercostal muscles	Medial borders of the coracoid process of the scapula	Protraction (abduction) of scapula; downward rotation of scapula; active in forced breathing	Lateral and medial pectoral nerves (C5, 6, 7, 8, and T1)
Subclavius	Superior surface of the junction of first rib and its cartilage	Groove on the inferior surface of the middle of the clavicle	Downward rotation of clavicle; stabilizes the clavicle	Nerve to the subclavius (C5 and C6)

Shoulder or Glenohumeral Joint

FIGURE 4.22

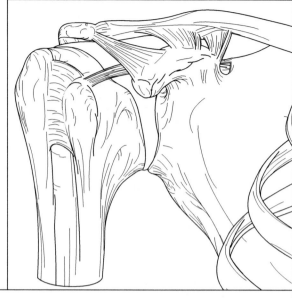

Type of Joint

Articulations

Ligaments

FIGURE 4.23

Movements of the Shoulder Joint

Flexion

Extension

Abduction

Adduction

Medial rotation

Lateral rotation

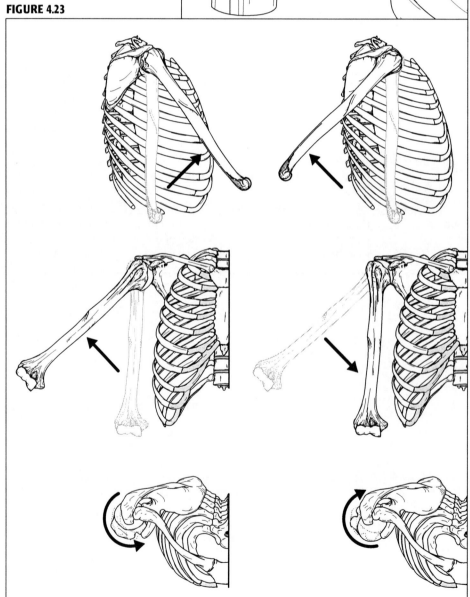

Illustrations by Jamey Garbett. © 2003 Mark Nielsen.

Muscles of the Shoulder Joint

The various muscles that cross the shoulder joint, which generate movements of the humerus, can be subdivided into four groups: rotator cuff muscle group, shoulder cap, intertubercular muscle group, and brachial muscles. Some muscles of the brachial group also cross the elbow joint.

Rotator Cuff Muscles
Supraspinatus Muscle
*supra- = superior or above + **spinatus** = referring to the spine of the scapula*

Origin

Insertion

Function

Structure and relationships

FIGURE 4.24

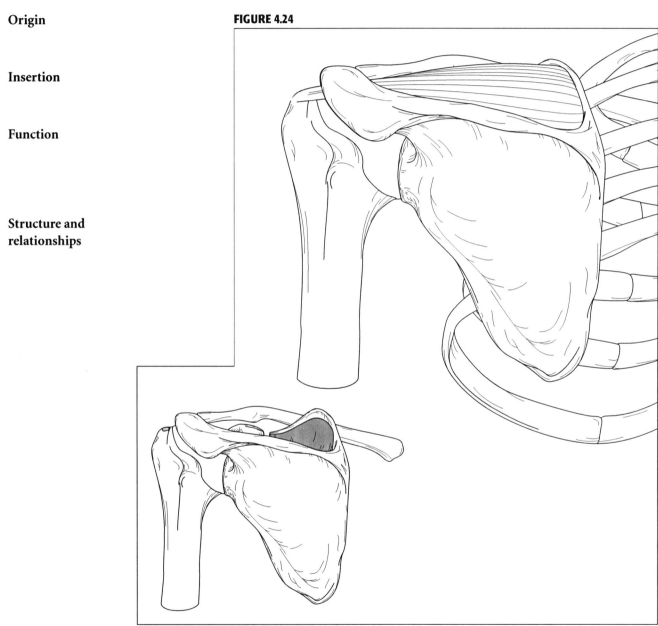

Illustration by Jamey Garbett. © 2003 Mark Nielsen.

Infraspinatus Muscle

infra- = *inferior or below* + *spinatus* = *referring to the spine of the scapula*

Origin

Insertion

Function

Structure and relationships

FIGURE 4.25

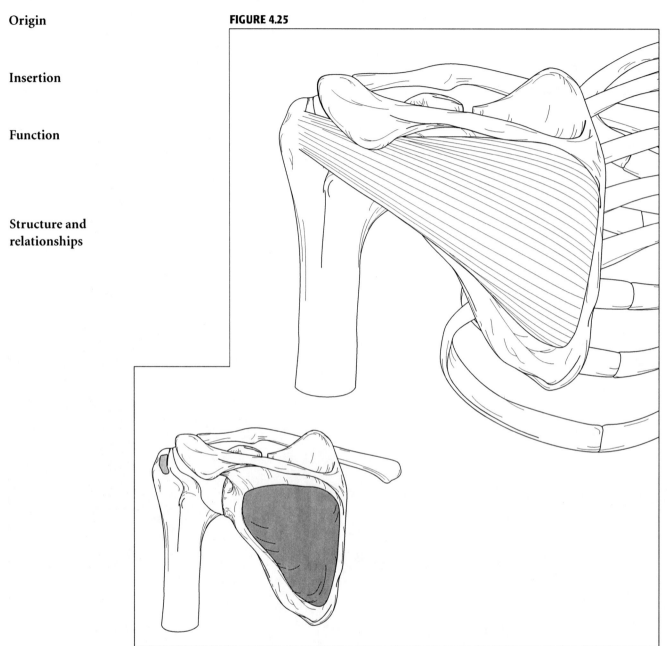

Illustration by Jamey Garbett. © 2003 Mark Nielsen.

Teres Minor Muscle

teres = round, minor = lesser

Origin

Insertion

Function

Structure and relationships

FIGURE 4.26

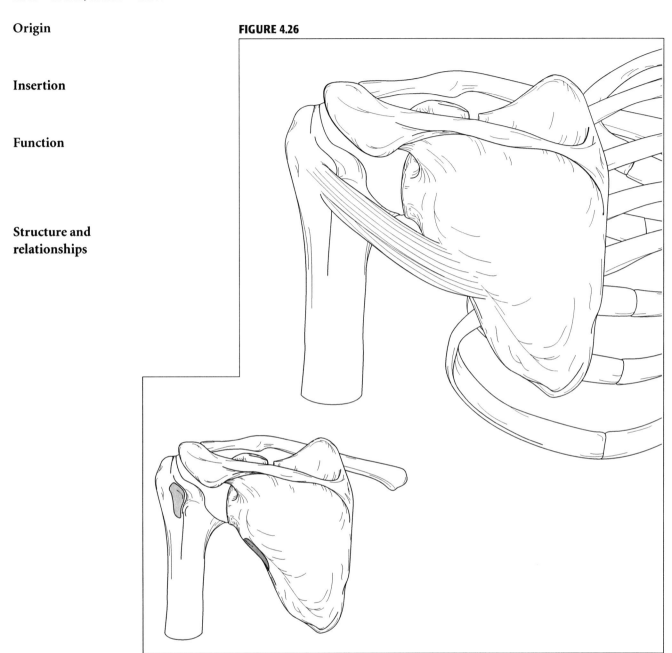

Illustration by Jamey Garbett. © 2003 Mark Nielsen.

Subscapularis Muscle

sub- = *beneath,* ***scapularis*** = *referring to the scapula*

Origin

Insertion

Function

**Structure and
relationships**

FIGURE 4.27

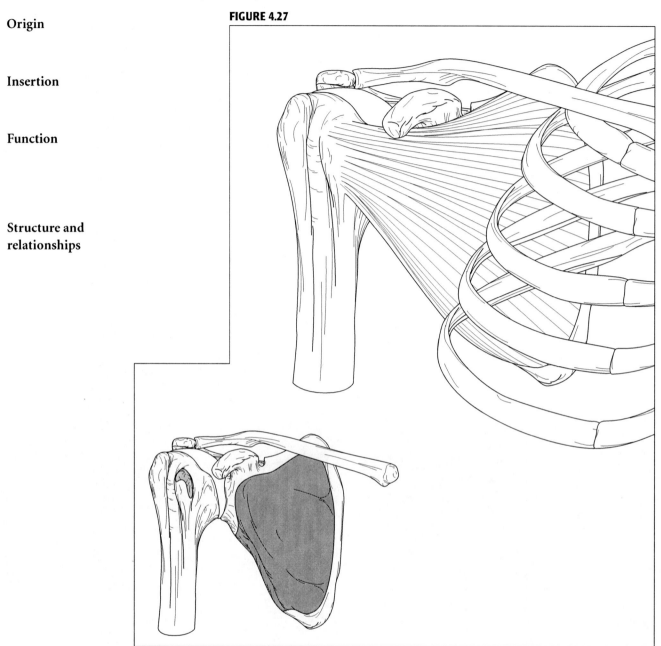

Illustration by Jamey Garbett. © 2003 Mark Nielsen.

Summary of the Rotator Cuff Muscles

Common attachments

Common actions at joint

FIGURE 4.28

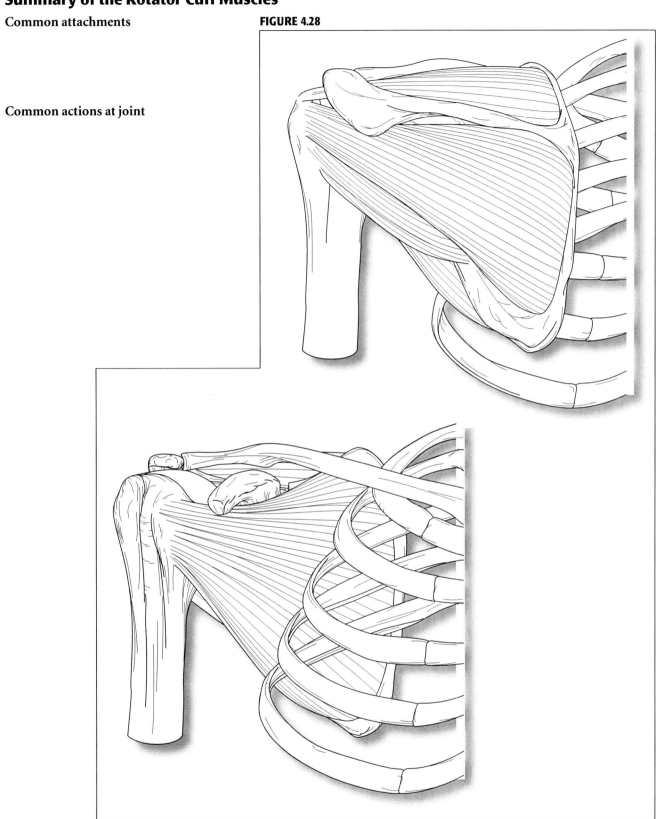

Illustration by Jamey Garbett. © 2003 Mark Nielsen.

Practical applications

FIGURE 4.29

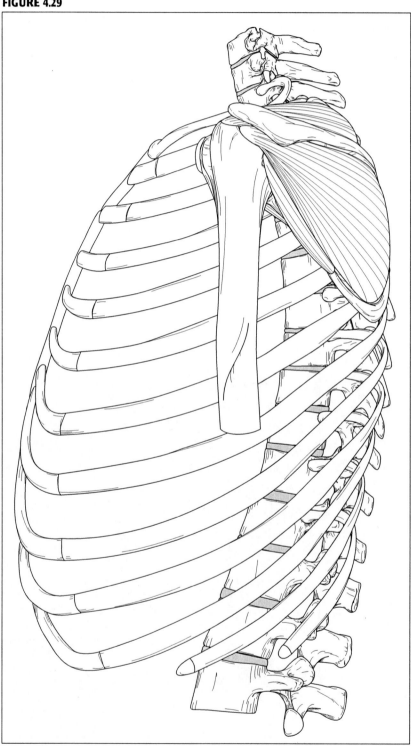

Illustration by Jamey Garbett. © 2003 Mark Nielsen.

Shoulder Cap

Deltoid Muscle

delta = the greek letter (triangle) + *-eidos* = shape

Origin

Insertion

Function

Structure and relationships

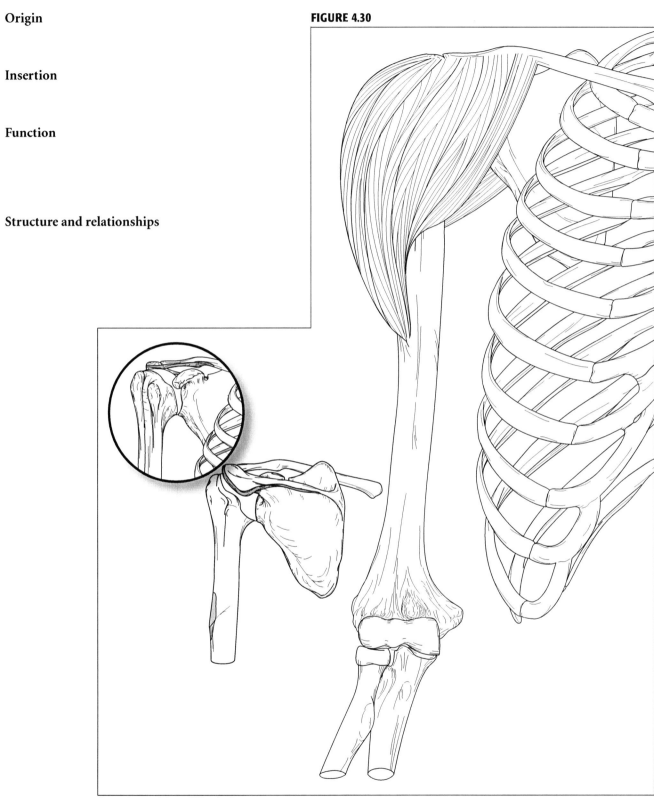

FIGURE 4.30

Illustration by Jamey Garbett. © 2003 Mark Nielsen.

Practical Applications
Raising Your Hand

What muscles come into play?

FIGURE 4.31

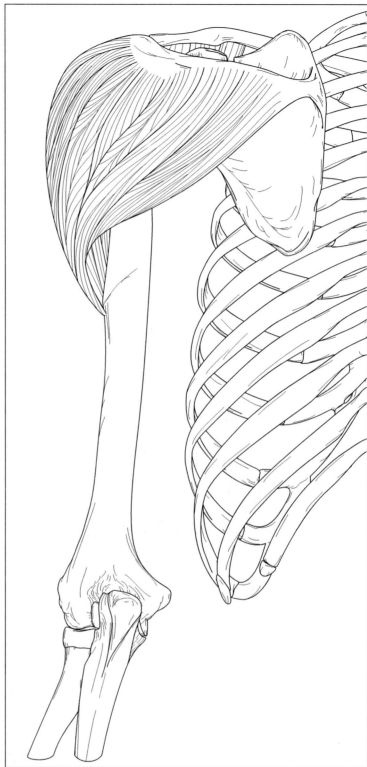

Illustration by Jamey Garbett. © 2003 Mark Nielsen.

Intertubercular Groove Muscles
Pectoralis Major Muscle

pectus = the breast, major = greater

Origin

Insertion

Function

Structure and relationships

FIGURE 4.32

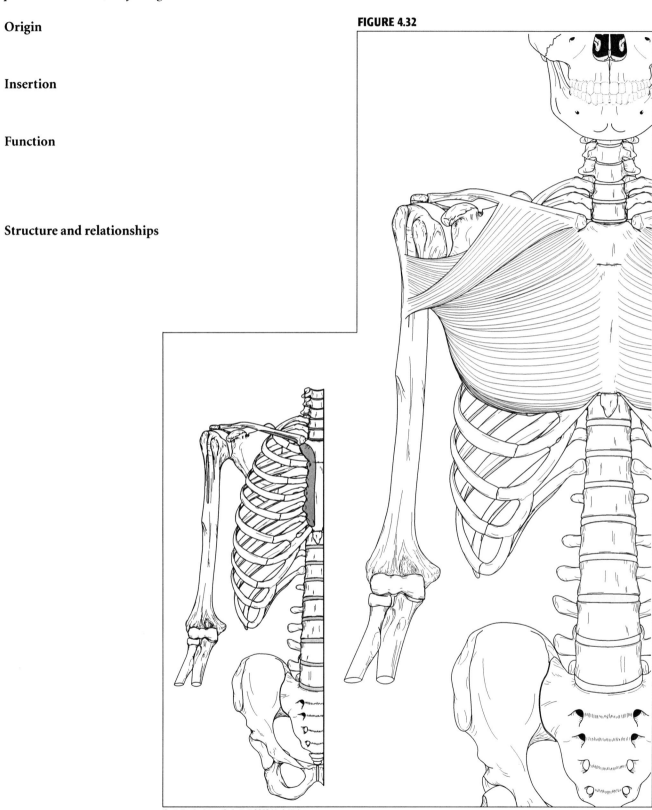

Illustration by Jamey Garbett. © 2003 Mark Nielsen.

Teres Major Muscle

teres = round, major = greater

Origin

Insertion

Function

Structure and relationships

FIGURE 4.33

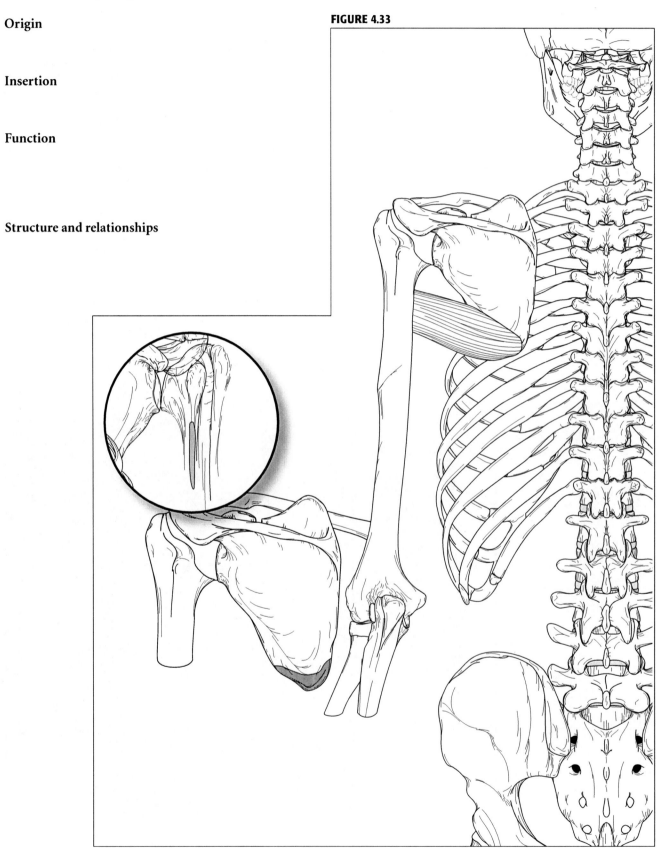

Illustration by Jamey Garbett. © 2003 Mark Nielsen.

Latissimus Dorsi Muscle
latus = broad, dorsum = back

Origin

Insertion

Function

Structure and relationships

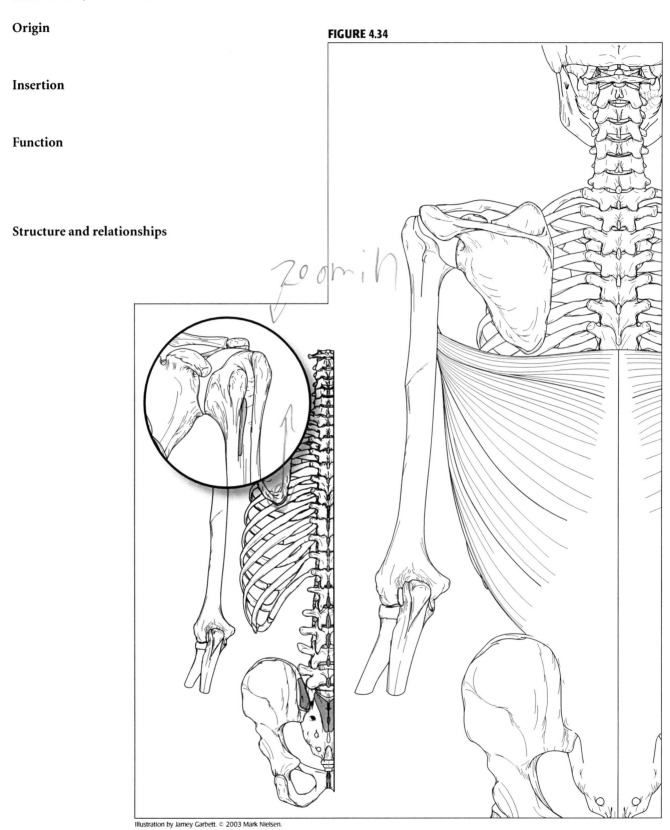

FIGURE 4.34

zoomih

Summary of the Intertubercular Groove Muscles

Common attachments

Common actions at joint

FIGURE 4.35

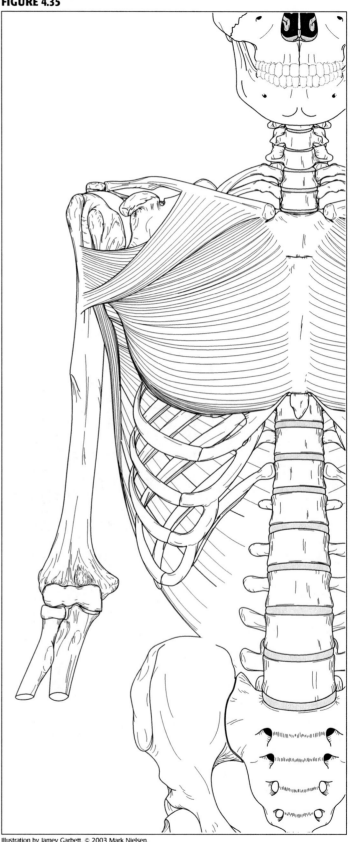

Illustration by Jamey Garbett. © 2003 Mark Nielsen.

Practical applications

FIGURE 4.36

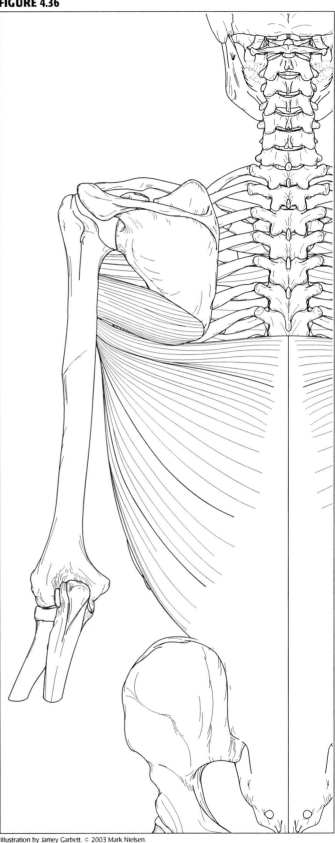

Illustration by Jamey Garbett. © 2003 Mark Nielsen.

Summary of the Shoulder Joint Muscles

FIGURE 4.37

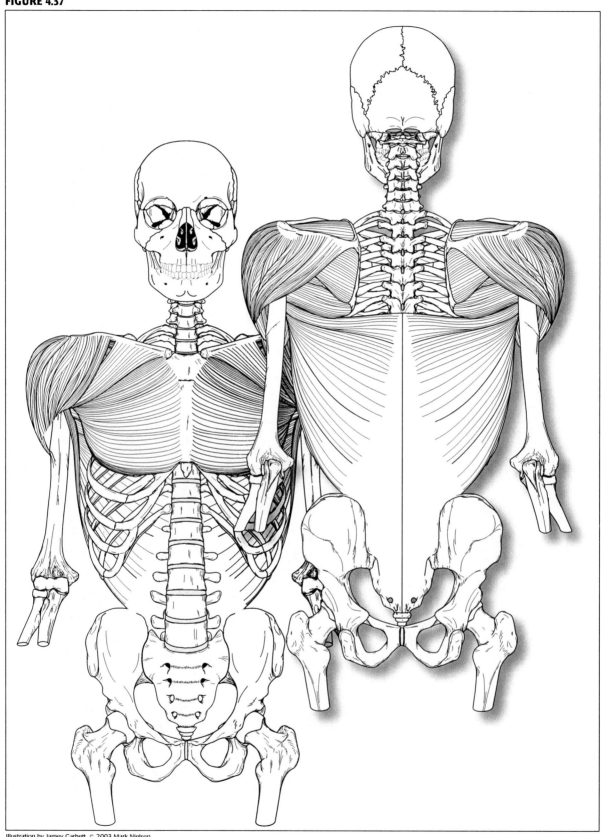

Illustration by Jamey Garbett. © 2003 Mark Nielsen.

Muscles of the Shoulder Joint

Muscle	Origin	Insertion	Function	Innervation
Supraspinatus	Supraspinous fossa of scapula and the overlying supraspinous fascia	Upper aspect of greater tubercle of humerus and articular capsule of the shoulder joint	Abduction of shoulder; stabilizes shoulder joint; supports downward pull of a loaded arm	Suprascapular nerve (C5 and C6)
Infraspinatus	Medial part of infraspinous fossa of scapula and from tendinous septa from ridges on infraspinous fossa more laterally	Middle aspect of greater tubercle of humerus	Lateral rotation of shoulder; stabilizes shoulder joint; holds humeral head down during initial abduction of humerus	Suprascapular nerve (C5 and C6)
Teres minor	Dorsal surface of the upper lateral margin of the scapula	Lowest aspect of greater tubercle of humerus and to humeral shaft just below the tubercle; inferior part of articular capsule of shoulder joint	Lateral rotation of shoulder; stabilizes shoulder joint; holds humeral head down during initial abduction of humerus	Axillary nerve (C5 and C6)
Subscapularis	Surface of the subscapular fossa and tendinous intermuscular septa that attach to the boney ridges of the subscapular fossa	Lesser tubercle of humerus and the articular capsule of the shoulder joint	Medial rotation of shoulder; stabilizes shoulder joint; holds humeral head down during initial abduction of humerus	Upper and lower subscapular nerves (C5 and C6)
Deltoid	Anterior superior border of lateral third of clavicle, lateral and superior parts of the acromion, inferior edge of the spine of the scapula	Deltoid tuberosity of humerus and brachial fascia	Flexion and weak medial rotation of the shoulder; abduction of shoulder; extension and weak lateral rotation of the shoulder	Axillary nerve (C5 and C6)
Pectoralis major	Anterior surface of the sternal half of clavicle, anterior surface of sternum, anterior surface of costal cartilages 1 to 7, and the aponeurosis of the external oblique muscle	Lateral ridge of the intertubercular groove via anterior and posterior laminae, the inferior sternal fibers attaching superiorly on intertubercular groove	Adduction of shoulder; medial rotation of shoulder; flexion of an extended shoulder; when arm is anchored above the trunk it pulls the trunk upward	Lateral and medial pectoral nerves (C5, 6, 7, 8, and T1)
Teres major	Flattened area on the dorsal surface of the inferior angle of the scapula	Medial ridge of the intertubercular groove	Adduction of shoulder; medial rotation of shoulder; extension of a flexed shoulder	Lower subscapular nerve (C5, C6, and C7)
Latissimus dorsi	Spinous processes from 7th to 12th thoracic vertebrae; from spinous processes of lumbar and sacral vertebrae via the thoracolumbar fascia; posterior part of the iliac crest; from outside surface of last three ribs	Floor of the intertubercular groove at a more proximal location than the teres major	Adduction of shoulder; medial rotation of shoulder; extension of a flexed shoulder; when arm is anchored above the trunk it pulls the trunk upward	Thoracodorsal nerve (C6, C7, and C8)

Elbow Joint

Type
Humero-ulnar joint

Humeroradial joint

Proximal radio-ulnar joint

Articulations
Humero-ulnar joint

Humeroradial joint

Proximal radio-ulnar joint

Ligaments
Ulnar collateral ligament

Radial collateral ligament

Annular ligament

FIGURE 4.38

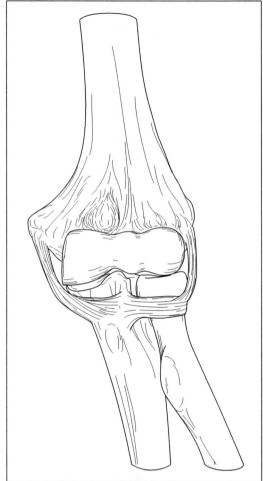

Illustration by Jamey Garbett. © 2003 Mark Nielsen.

Movements of the Elbow Joint

Flexion

Extension

FIGURE 4.39

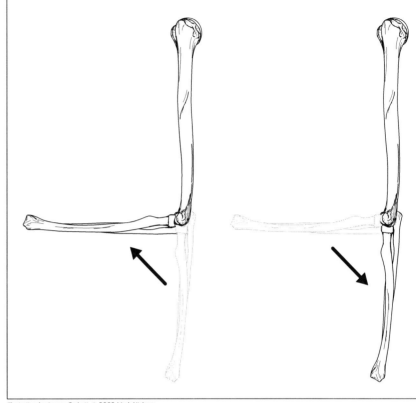

Illustration by Jamey Garbett. © 2003 Mark Nielsen.

Supination

Pronation

FIGURE 4.40

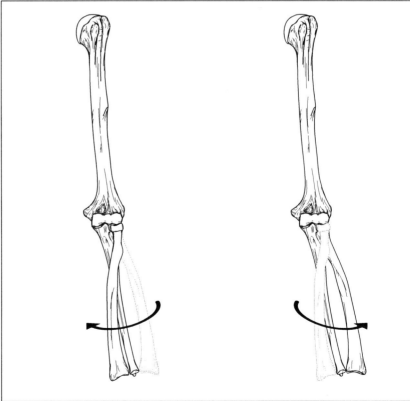

Illustration by Jamey Garbett. © 2003 Mark Nielsen.

Brachial Muscles

Anterior Brachial Compartment Muscles

Coracobrachialis Muscle

coraco- = referring to the coracoid process + brachium = arm

Origin

Insertion

Function

Structure and relationships

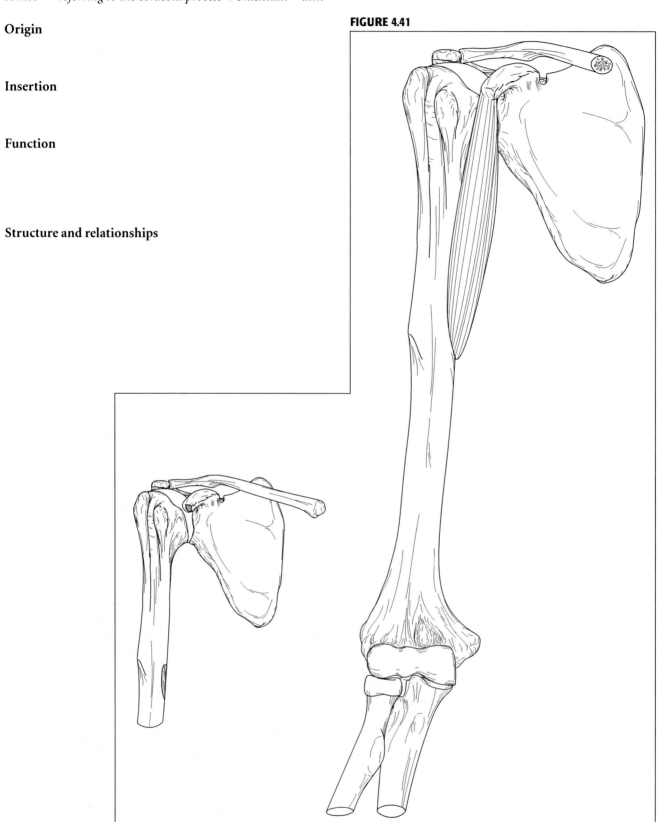

FIGURE 4.41

Illustration by Jamey Garbett. © 2003 Mark Nielsen.

Brachialis Muscle

brachium = arm

Origin

Insertion

Function

Structure and relationships

FIGURE 4.42

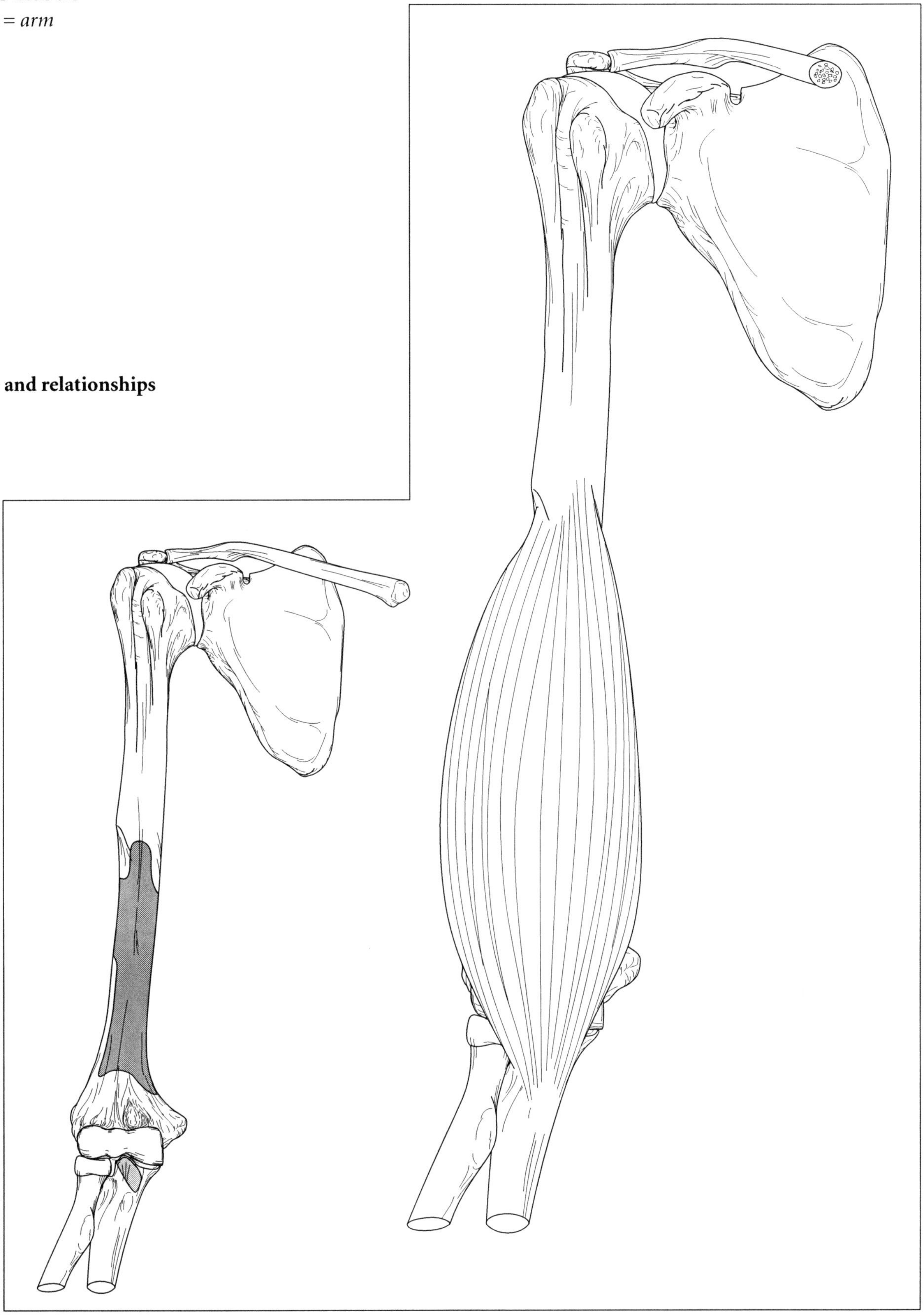

Illustration by Jamey Garbett. © 2003 Mark Nielsen.

Biceps Brachii Muscle

biceps = two heads, brachium = arm

Origin

Insertion

Function

Structure and relationships

FIGURE 4.43

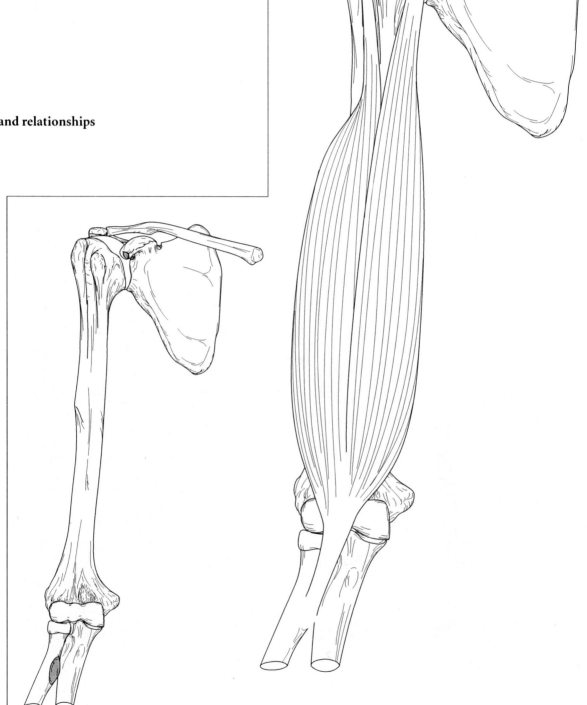

Illustration by Jamey Garbett. © 2003 Mark Nielsen.

Summary of the Anterior Brachial Muscles

Common attachments

Common actions at joints

Practical applications

FIGURE 4.44

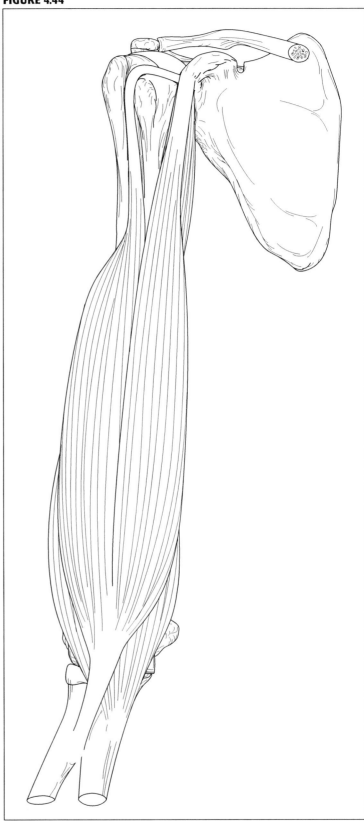

Illustration by Jamey Garbett. © 2003 Mark Nielsen.

Posterior Brachial Compartment Muscles
Triceps Brachii Muscle
triceps = three heads, brachium = arm

Origin

Insertion

Function

Structure and relationships

FIGURE 4.45

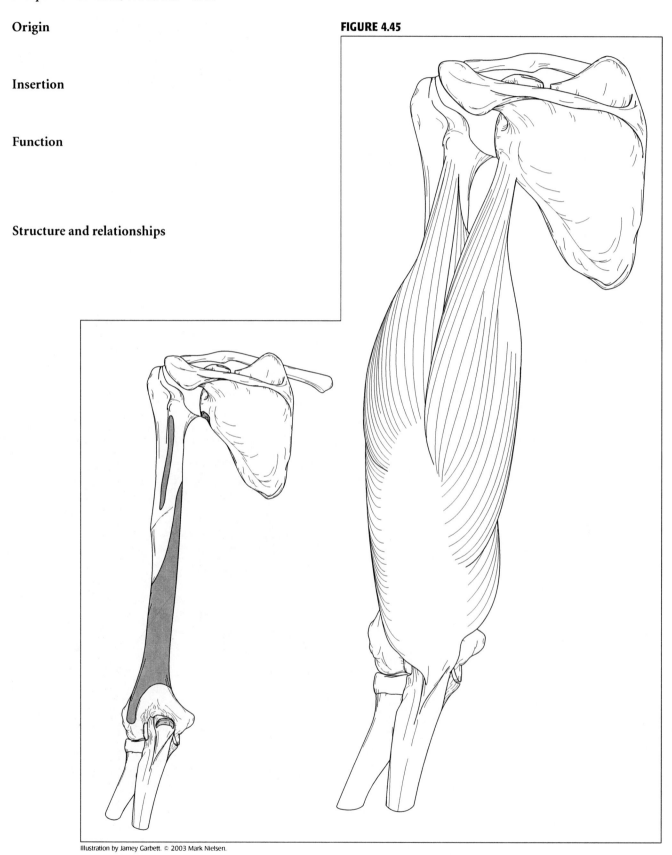

Illustration by Jamey Garbett. © 2003 Mark Nielsen.

Muscles of the Brachium

Muscle	Origin	Insertion	Function	Nerve Supply
Coracobrachialis	Apex of the coracoid process of the scapula	Medial shaft of humerus between the brachialis and triceps brachii	Flexion of shoulder, most effective when the arm is extended	Musculocutaneous nerve (C5, C6, and C7)
Brachialis	Anterior surface of humerus from just proximal to the deltoid attachment to just proximal to trochlea and capitulum	Anterior aspect of coronoid process and ulnar tuberosity	Flexion of elbow	Musculocutaneous nerve (C5 and C6)
Biceps brachii	Apex of coracoid process (short head) and supraglenoid tubercle of scapula (long head)	Radial tuberosity and deep fascia of antebrachial flexor muscles via bicipital aponeurosis	Major supinator; flexes elbow (most powerfully when the forearm is supinated); assists with flexion of the shoulder joint	Musculocutaneous nerve (C5 and C6)
Triceps brachii	Infraglenoid tubercle of scapula and joint capsule of the shoulder (long head), oblique ridge in the upper half of the posterior humeral shaft (lateral head), posterior and medial surfaces of the humeral shaft below the point of insertion of the teres major to just proximal to the trochlea (medial head)	Upper surface of the olecranon process of ulna and the antebrachial fascia	Extension of the elbow joint; assist with extension of the shoulder especially when the arm is flexed; long head helps stabilize the bottom of the shoulder joint	Radial nerve (C6, C7, and C8)

Practical applications

Muscle Topography in the Shoulder and Brachium
Muscle Views

FIGURE 4.46

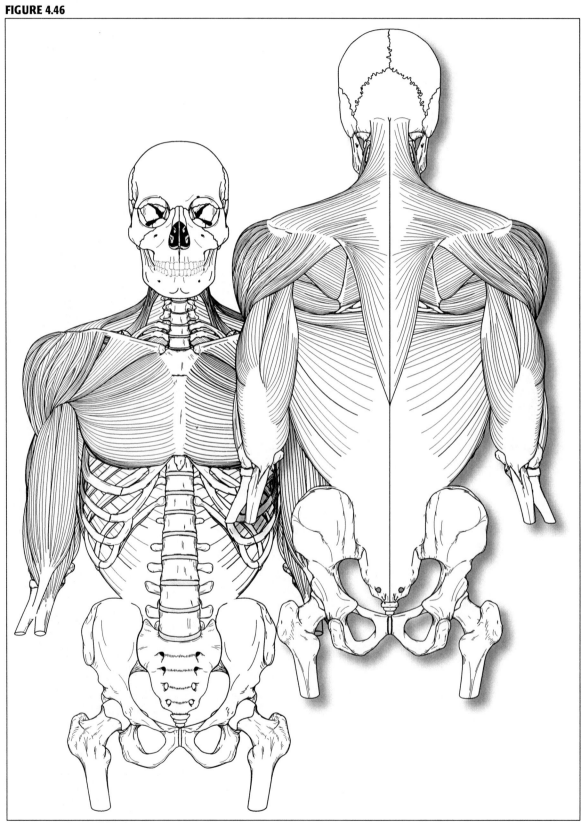

Illustration by Jamey Garbett. © 2003 Mark Nielsen.

Wrist Joint

Type of Joint

Articulations

Ligaments

Movements of the Wrist Joint

Flexion

Extension

Abduction or Radial Deviation

Adduction or Ulnar Deviation

FIGURE 4.47

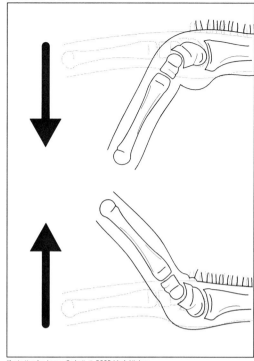

Illustration by Jamey Garbett. © 2003 Mark Nielsen.

FIGURE 4.48

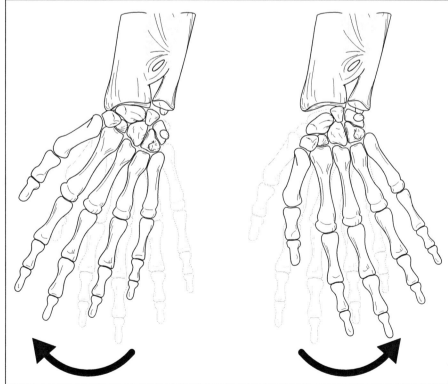

Illustration by Jamey Garbett. © 2003 Mark Nielsen.

Finger Joints

Metacarpophalangeal (MP) Joints

meta = beyond + karpos = wrist + phalange = referring to the fingers

Movements

Extension

Flexion

Abduction

Adduction

FIGURE 4.49

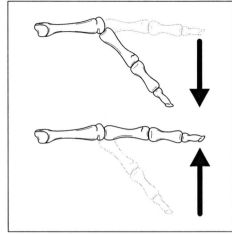

Illustration by Jamey Garbett. © 2003 Mark Nielsen.

FIGURE 4.50

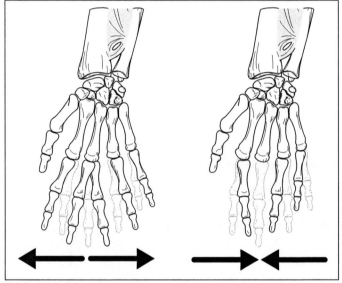

Illustration by Jamey Garbett. © 2003 Mark Nielsen.

Interphalangeal (IP) Joints

inter- = beyond + phalange = referring to the fingers

Movements

Flexion

Extension

FIGURE 4.51

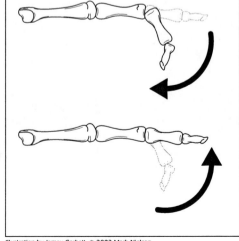

Illustration by Jamey Garbett. © 2003 Mark Nielsen.

Thumb Joints
Carpometacarpal Joint Movements
Flexion

Extension

Abduction

Adduction

Opposition

Reposition

FIGURE 4.52

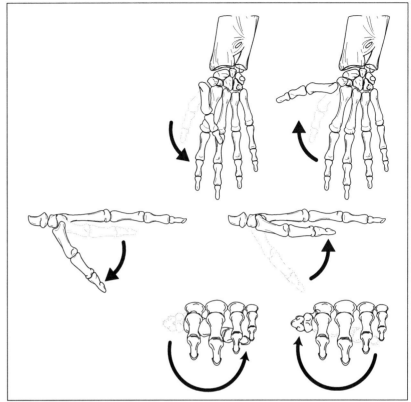

Illustration by Jamey Garbett. © 2003 Mark Nielsen.

Metacarpophalangeal (MP) Joint Movements
Flexion

Extension

FIGURE 4.53

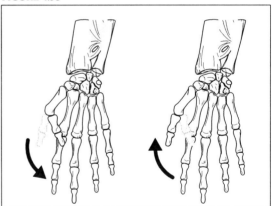

Illustration by Jamey Garbett. © 2003 Mark Nielsen.

Interphalangeal (IP) Joint Movements
Flexion

Extension

FIGURE 4.54

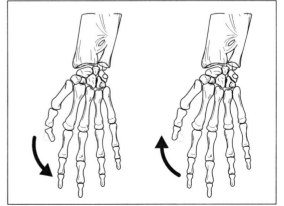

Illustration by Jamey Garbett. © 2003 Mark Nielsen.

Antebrachial Muscles

Anterior Antebrachial Compartment Muscles
Superficial Group

Pronator Teres Muscle
*pronator = one that pronates, **teres** = round*

Structure and relationships

Function

Palmaris Longus Muscle

*palmaris = referring to the palm of the hand, **longus** = long*

Structure and relationships

Function

FIGURE 4.57

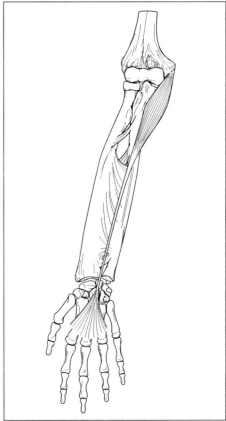

Illustration by Jamey Garbett. © 2003 Mark Nielsen.

Flexor Carpi Ulnaris Muscle

*flexor = one that flexes, **carpus** = wrist, **ulnaris** = referring to the ulna*

Structure and relationships

Function

FIGURE 4.58

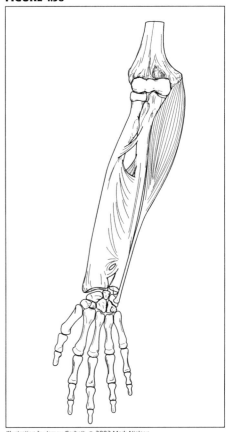

Illustration by Jamey Garbett. © 2003 Mark Nielsen.

Flexor Digitorum Superficialis Muscle

flexor = one that flexes, *digitorum* = referring to the fingers, *superficialis* = outer surface

This muscle is deep to the other superficial muscles and forms somewhat of an intermediate plane of muscle between the four preceding superficial muscles and the deep group.

Structure and relationships

Function

FIGURE 4.59

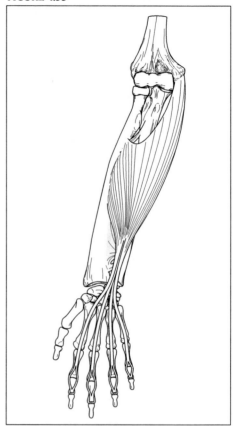

Illustration by Jamey Garbett. © 2003 Mark Nielsen.

Deep Group
Flexor Digitorum Profundus Muscle

flexor = one that flexes, *digitorum* = referring to the fingers, *profundus* = deeper

Structure and relationships

Function

FIGURE 4.60

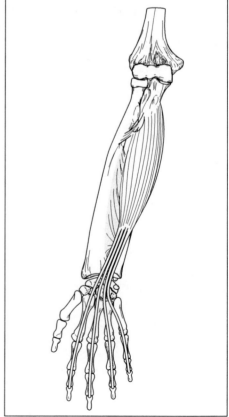

Illustration by Jamey Garbett. © 2003 Mark Nielsen.

Flexor Pollicis Longus Muscle

flexor = one that flexes, ***pollex*** = thumb, ***longus*** = long

Structure and relationships

Function

Illustration by Jamey Garbett. © 2003 Mark Nielsen.

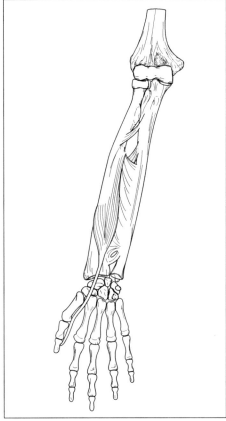

FIGURE 4.61

Pronator Quadratus Muscle

pronator = one that pronates, ***quadratus*** = four-sided

Structure and relationships

Function

Illustration by Jamey Garbett. © 2003 Mark Nielsen.

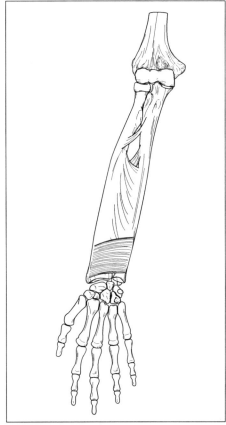

FIGURE 4.62

Summary of the Anterior Antebrachial Compartment Muscles

Common attachments

Common actions at joints

Practical applications

FIGURE 4.63

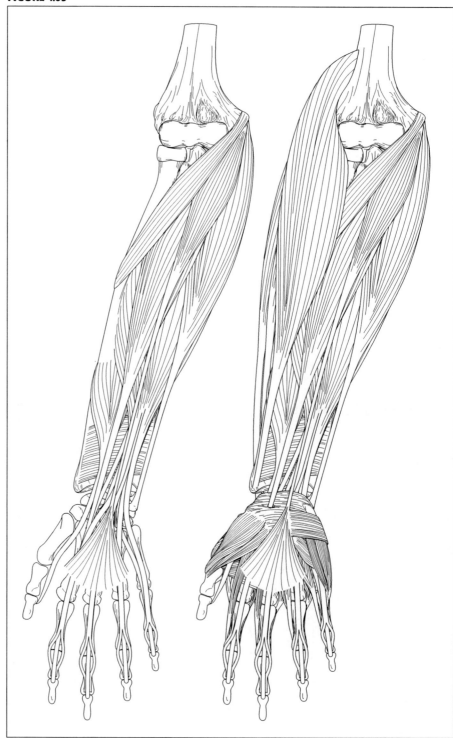

Illustration by Jamey Garbett. © 2003 Mark Nielsen.

Muscles of the Anterior Antebrachium

Muscle	Origin	Insertion	Function	Innervation
Pronator teres	Slightly proximal to the medial epicondyle of humerus and from the common flexor tendon (humeral head); medial aspect of the coronoid process of the ulna (ulnar head)	Lateral surface of radial shaft near its midpoint	Pronation (it is the weaker of the two pronator muscles, only active during rapid, forcible pronation); weak flexor of elbow joint	Median nerve (C6 and C7)
Flexor carpi radialis	Medial epicondyle of humerus via the common flexor tendon	Anterior surface of the base of 2nd metacarpal bone; small tendinous slip to the 3rd metacarpal base	Flexion of wrist; abduction (radial deviation) of wrist	Median nerve (C6 and C7)
Palmaris longus	Medial epicondyle of humerus via the common flexor tendon	Flexor retinaculum and via palmar aponeurosis to skin and fascia of palm and digits	Main function is to anchor the skin and fascia of the hand and keep it from sliding towards the digits; assists flexion of the wrist	Median nerve (C7 and C8)
Flexor carpi ulnaris	Medial epicondyle of humerus via the common flexor tendon (humeral head); medial side of the olecranon and the proximal half of the posterior ulnar shaft (ulnar head)	Pisiform bone, hamate bone, anterior base of 5th metacarpal, and the flexor retinaculum	Flexion of wrist; adduction (ulnar deviation) of wrist	Ulnar nerve (C7, C8, and T1)
Flexor digitorum superficialis	Medial epicondyle of humerus via the common flexor tendon, anterior surface of ulnar collateral ligament, medial surface of the coronoid process (humeroulnar head); anterior surface of radius from radial tuberosity to mid-radius (radial head)	Individual tendons split and wrap around deep tendons before reuniting to attach to anterior surface of the middle phalanges of the four fingers (independent tendons to all fingers)	Flexion of wrist and digits (except last digital joint); tendon arrangement allows it to flex one proximal interphalangeal joint at a time	Median nerve (C8 and T1)
Flexor digitorum profundus	Medial side of the coronoid process, posterior ulnar border and anterior and medial surfaces of the ulna along the length of the ulnar shaft, ulnar side of the interosseous membrane	Anterior surface of base of distal phalanges of fingers (muscle group common to middle, ring, and little fingers, with numerous interconnections between the three tendons)	Flexion of wrist and digits (all joints)	Median and ulnar nerves (C8 and T1)
Flexor pollicis longus	Anterior surface of radius from the base of the radial tuberosity to the pronator quadratus, radial side of the interosseous membrane	Anterior surface of the base of the distal phalanx of the thumb	Flexes the phalanges of the thumb; weakly assists flexors of wrist	Median nerve (C7 and C8)
Pronator quadratus	Oblique ridge on anterior surface of the distal ulna	Distal quarter of the radius on the anterolateral surface and into the area just above the ulnar notch of the radius	Major pronator (always active in pronation); stabilizes separation of the radius and ulna when hand is forced into wrist joint	Median nerve (C7 and C8)

Posterior Antebrachial Compartment Muscles
Lateral Group
Brachioradialis Muscle
brachium = arm, radialis = referring to the radius

Structure and relationships

Function

FIGURE 4.64

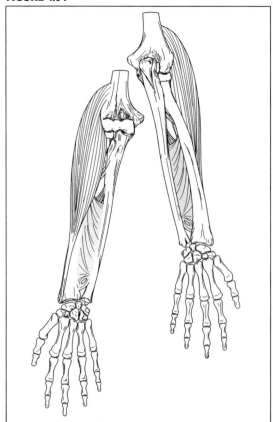

Illustration by Jamey Garbett. © 2003 Mark Nielsen.

Extensor Carpi Radialis Longus Muscle
extensor = one that extends, carpus = wrist,
radialis = referring to the radius, longus = long

Structure and relationships

Function

FIGURE 4.65

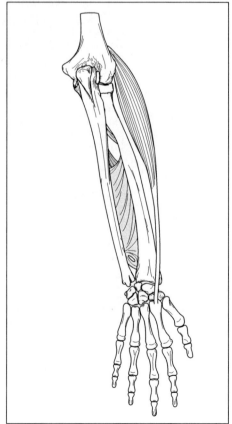

Illustration by Jamey Garbett. © 2003 Mark Nielsen.

Extensor Carpi Radialis Brevis Muscle

extensor = one that extends, carpus = wrist, radialis = referring to the radius, brevis = short

Structure and relationships

Function

FIGURE 4.66

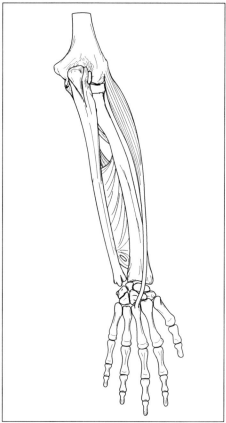

Illustration by Jamey Garbett. © 2003 Mark Nielsen.

Extensor Digitorum Muscle

extensor = one that extends, digitorum = referring to the fingers

Structure and relationships

Function

FIGURE 4.67

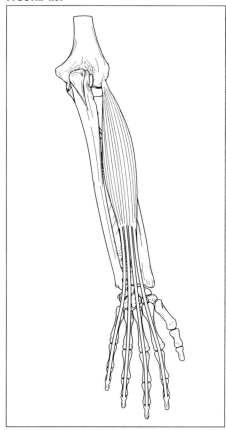

Illustration by Jamey Garbett. © 2003 Mark Nielsen.

Extensor Digiti Minimi Muscle

*extensor = one that extends, **digiti minimi** = little finger*

Structure and relationships

Function

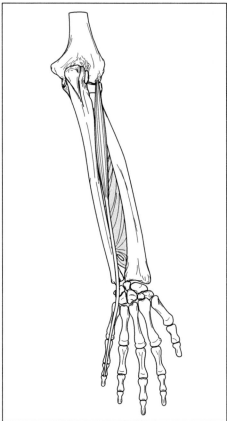

FIGURE 4.68

Illustration by Jamey Garbett. © 2003 Mark Nielsen.

Extensor Carpi Ulnaris Muscle

*extensor = one that extends, **carpus** = wrist, **ulnaris** = referring to the ulna*

Structure and relationships

Function

FIGURE 4.69

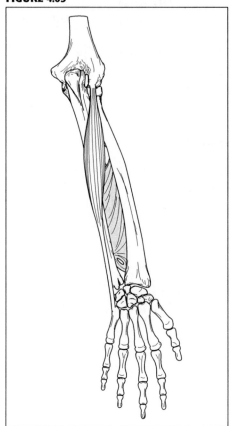

Illustration by Jamey Garbett. © 2003 Mark Nielsen.

Supinator Muscle

supinator = one that supinates

Structure and relationships

Function

FIGURE 4.70

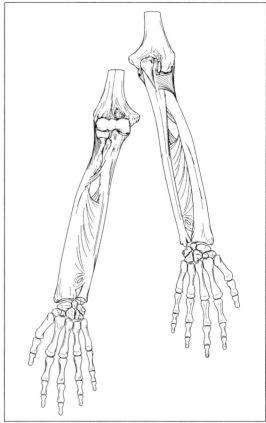

Illustration by Jamey Garbett. © 2003 Mark Nielsen.

Anconeus Muscle

ancon = bent, referring to the elbow

Structure and relationships

Function

FIGURE 4.71

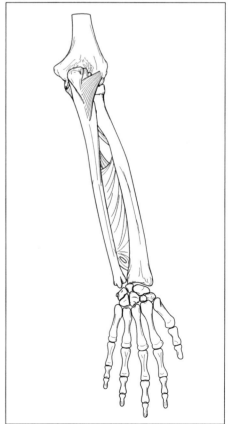

Illustration by Jamey Garbett. © 2003 Mark Nielsen.

Radial Group
Abductor Pollicis Longus Muscle
abductor = one that abducts, ***pollex*** = thumb, ***longus*** = long

Structure and relationships

Function

FIGURE 4.72

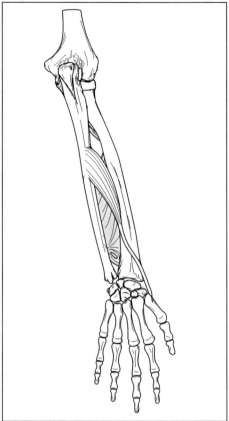

Illustration by Jamey Garbett. © 2003 Mark Nielsen.

Extensor Pollicis Longus Muscle
extensor = one that extends, ***pollex*** = thumb, ***longus*** = long

Structure and relationships

Function

FIGURE 4.73

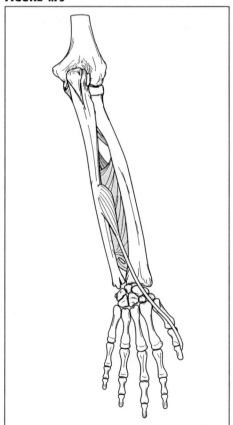

Illustration by Jamey Garbett. © 2003 Mark Nielsen.

Extensor Pollicis Brevis Muscle

extensor = one that extends, pollex = thumb, brevis = short

Structure and relationships

Function

FIGURE 4.74

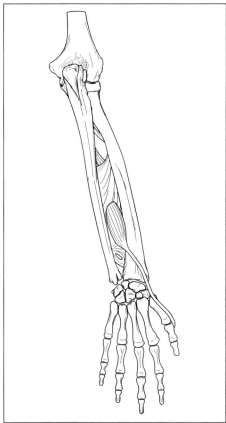

Illustration by Jamey Garbett. © 2003 Mark Nielsen.

Extensor Indicis Muscle

extensor = one that extends, indicis = referring to the index finger

Structure and relationships

Function

FIGURE 4.75

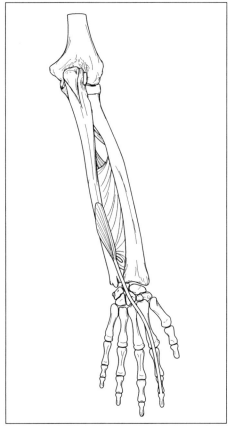

Illustration by Jamey Garbett. © 2003 Mark Nielsen.

Summary of the Posterior Antebrachial Compartment Muscles

Common attachments

Common actions at joints

Practical applications

FIGURE 4.76

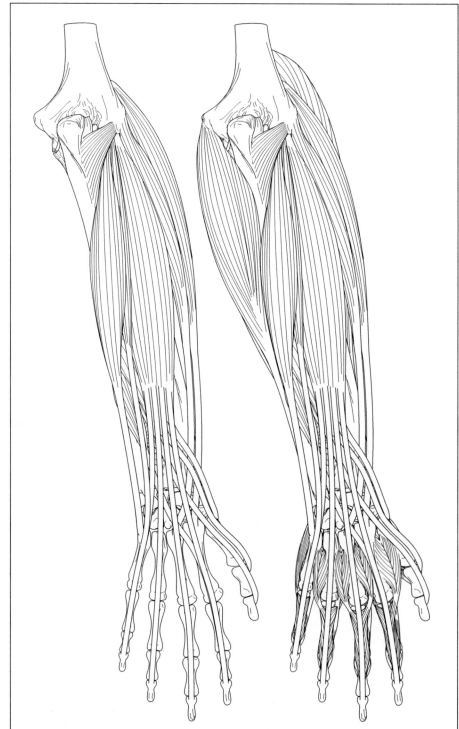

Illustration by Jamey Garbett. © 2003 Mark Nielsen.

Muscles of the Posterior Antebrachium

Muscle	Origin	Insertion	Function	Innervation
Brachioradialis	Proximal part of the lateral supracondylar ridge of the humerus and the lateral intermuscular septum	Distal end of lateral shaft of radius and base of radial styloid process	Flexes the elbow (best mechanical advantage when the forearm is semiprone); stabilizes the elbow joint during rapid elbow movements	Radial nerve (C5 and C6)
Extensor carpi radialis longus	Distal part of lateral supracondylar ridge of the humerus and lateral epicondyle via the common extensor tendon	Posterior surface at the base of 2nd metacarpal bone	Extension of the wrist; abducts (radial deviation) the wrist	Radial nerve (C6 and C7)
Extensor carpi radialis brevis	Lateral epicondyle of humerus via common extensor tendon and radial collateral ligament	Posterior surface at the base of the 3rd metacarpal bone	Extension of the wrist; abducts (radial deviation) the wrist	Radial nerve (C7 and C8)
Extensor digitorum	Lateral epicondyle of humerus via the common extensor tendon	Posterior surface of the phalanges via the extensor expansion	Extension of the wrist and all digital joints	Radial nerve (C7 and C8)
Extensor digiti minimi	Lateral epicondyle of humerus via a thin tendinous slip to common extensor tendon	Tendon divides into two distally and both tendons join the extensor expansion of the little finger with the extensor digitorum	Extension of all the joints of the little finger and assists wrist extensors	Radial nerve (C7 and C8)
Extensor carpi ulnaris	Lateral epicondyle of humerus via common extensor tendon and from the posterior border of the ulna	Medial side of the base of the 5th metacarpal bone	Extension of the wrist; adduction (ulnar deviation) of the wrist	Radial nerve (C7 and C8)
Anconeus	Posterior surface of the lateral epicondyle of the humerus	Lateral surface of the olecranon and the proximal shaft of the ulna	Assists triceps with extension of the elbow	Radial nerve (C6 and C7)
Supinator	Posterior surface of the lateral epicondyle of the humerus, radial collateral ligament, annular liga-ment, and the lateral edge of the proximal ulna	Lateral and anterior surfaces of the proximal third of the shaft of radius	Supination (active in all phases and ranges of supination, though not as powerful as the biceps)	Radial nerve (C6 and C7)
Abductor pollicis longus	Posterior surface of ulnar shaft in its middle portion, the ulnar side of the interosseus membrane, and the posterior side of the radius in the middle third	Splits into two tendons, one to the base of the first metacarpal bone on the radial (palmar) side, the other to the trapezium bone	Assists with abduction of the thumb; assists with extension of the carpometacarpal joint of the thumb	Radial nerve (C7 and C8)
Extensor pollicis longus	Lateral side of the posterior surface of the ulna distal to the abductor pollicis longus and the adjacent interosseous membrane	Posterior surface of the base of distal phalanx of the thumb	Extends the metacarpal bone and the proximal and distal phalanges of the thumb	Radial nerve (C7 and C8)
Extensor pollicis brevis	Posterior surface of the radius distal to the abductor pollicis and the adjacent interosseous membrane	Dorsolateral surface on the base of the proximal phalanx of the thumb	Extends the metacarpal bone and proximal phalanx of the thumb	Radial nerve (C7 and C8)
Extensor indicis	Posterior surface of ulna in lower half where it attaches distal to the extensor pollicis longus	Ulnar side of extensor digitorum tendon near head of metacarpal bone to join the extensor expansion	Assists with extension of the wrist and the index finger	Radial nerve (C7 and C8)

Upper Limb

Label the muscles shown on the following pages.

FIGURE 4.77

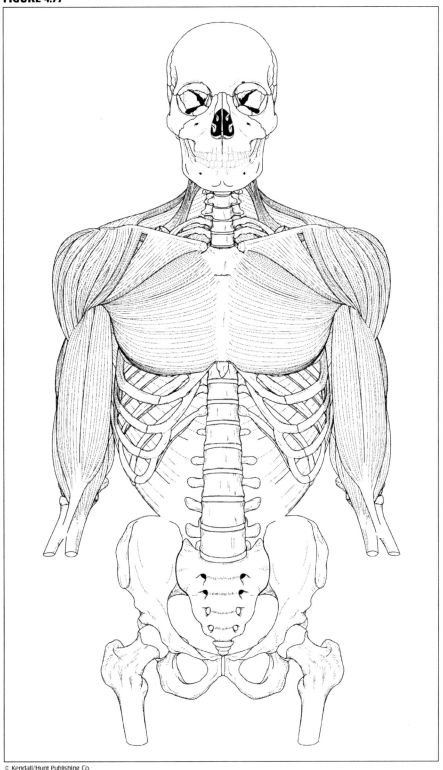

FIGURE 4.78

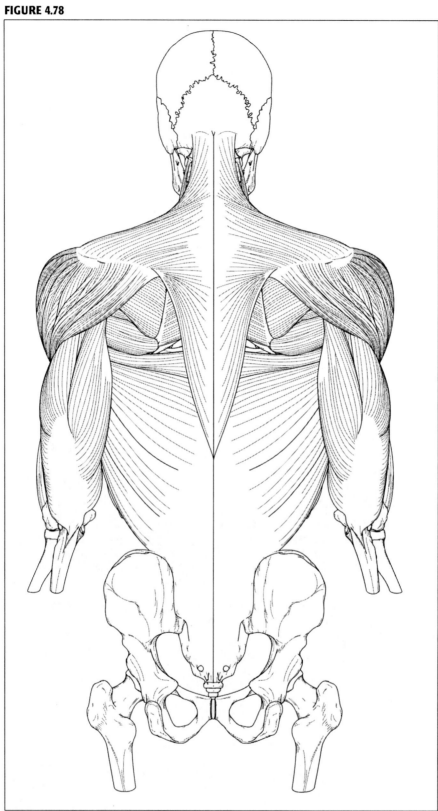

FIGURE 4.79

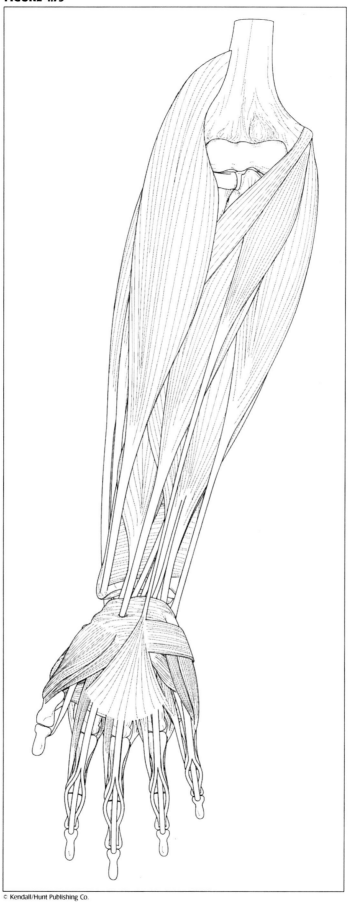

FIGURE 4.80

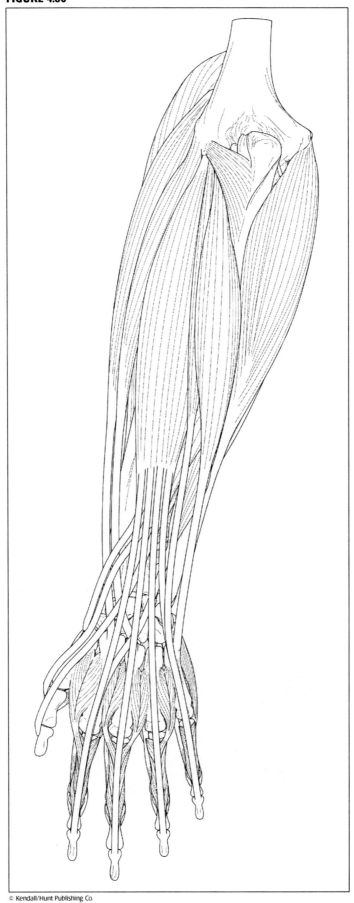

FIGURE 4.81

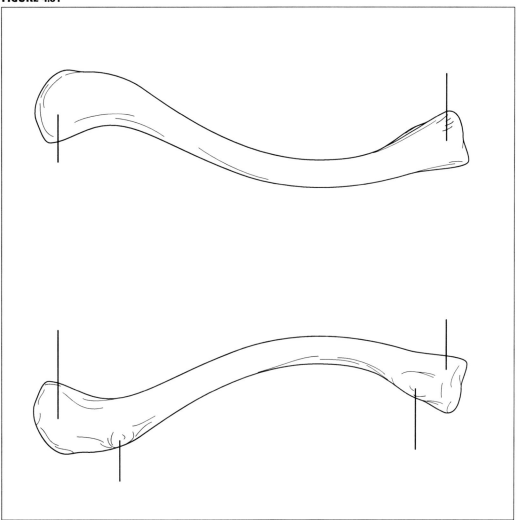

Illustration by Jamey Garbett. © 2003 Mark Nielsen.

FIGURE 4.82

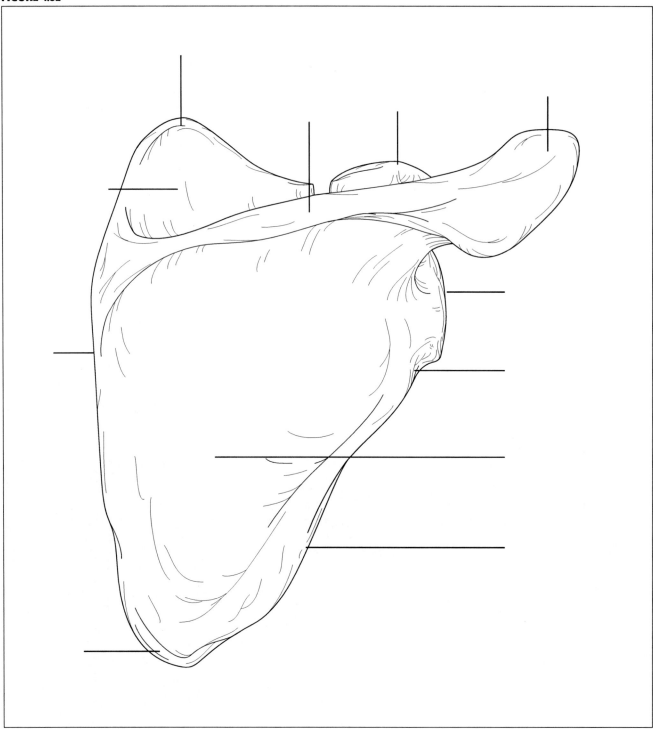

Illustration by Jamey Garbett. © 2003 Mark Nielsen.

150

FIGURE 4.83

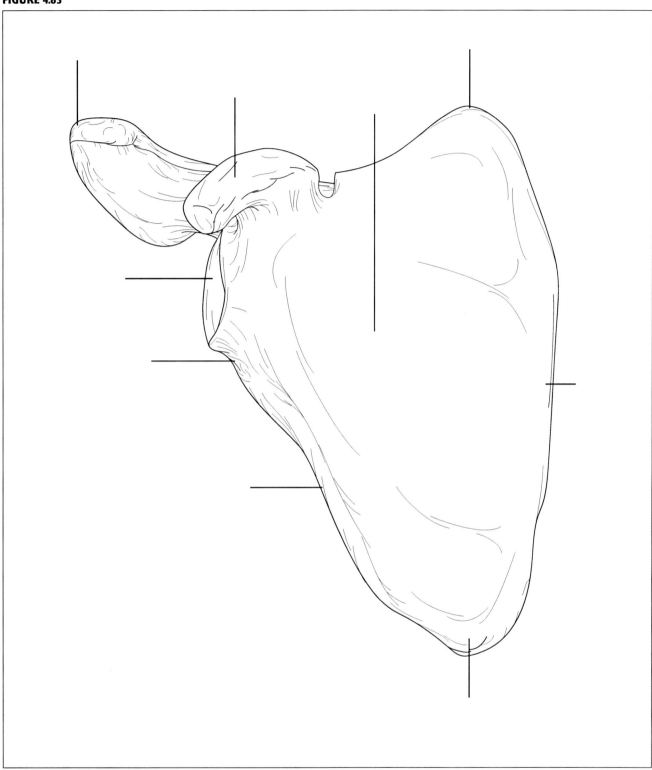

Illustration by Jamey Garbett. © 2003 Mark Nielsen.

151

FIGURE 4.84

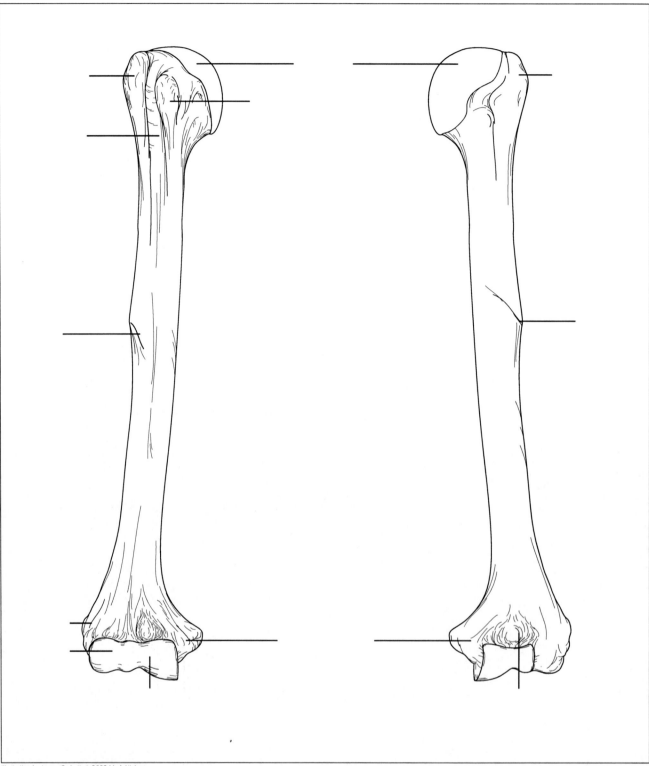

FIGURE 4.85

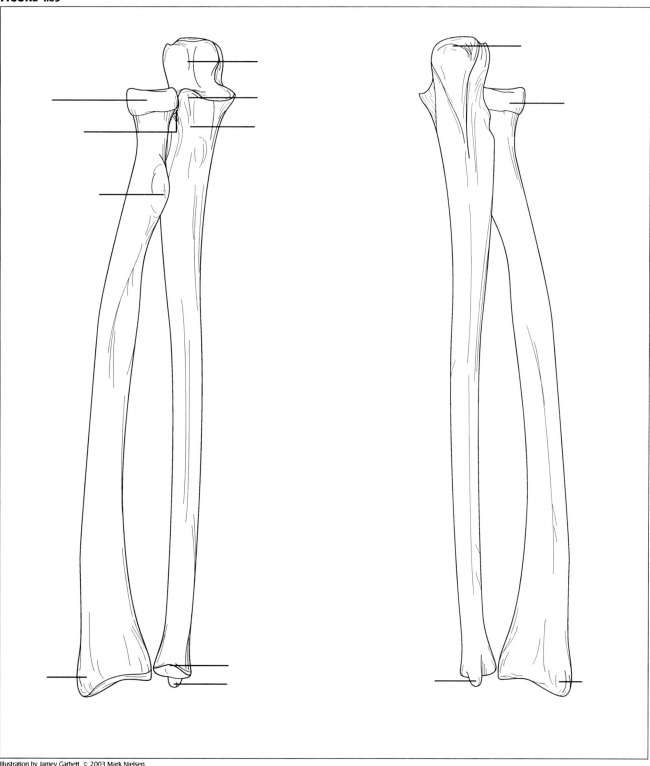

FIGURE 4.86

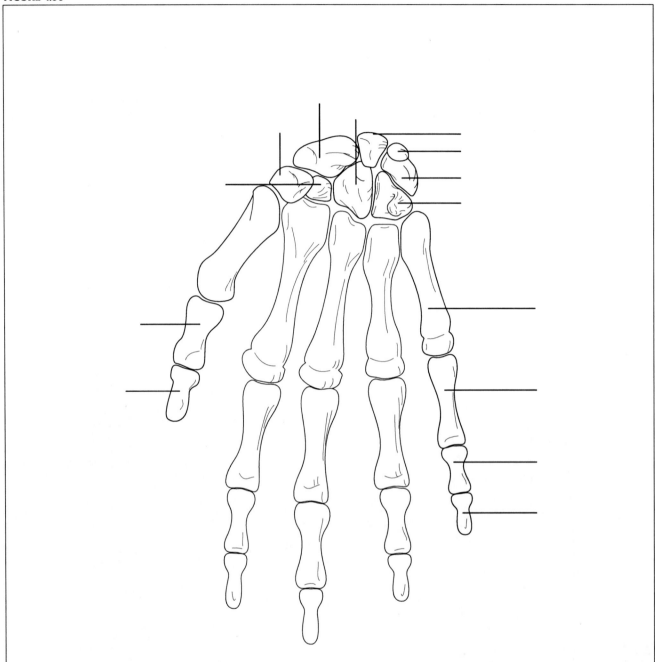

Illustration by Jamey Garbett. © 2003 Mark Nielsen.

Lower Limb

The lower limb bones include the phalanges, the metatarsals, the tarsals, the tibia, the fibula, and the femur. Additionally, the three bones of the pelvic girdle, the ilium, ischium, and pubis, are considered part of the lower limb. The pelvic girdle connects the lower limb to the axial skeleton at the sacroiliac joint. This joint is very strong and transfers the weight of the body to the lower limbs.

OBJECTIVES

☑ Identify the bones of the lower limb and the names of the joints where they articulate
☑ Know the various movements that can occur at each joint
☑ Identify the muscles of the lower limb and know their function

STRUCTURES YOU ARE RESPONSIBLE FOR IDENTIFYING

Bones and Landmarks
Coxa (innominate bone)
    Acetabulum
    Obturator foramen
Ilium
    Iliac crest
    Anterior superior iliac spine
    Iliac fossa
    Pelvic inlet
    Pelvic outlet
    True pelvis
    False pelvis
Ischium
    Ischial tuberosity
    Ischial spine
    Greater sciatic notch
Pubis
    Pubic tubercle
    Ramus
    Pectineal line
    Pubic symphysis
Femur
    Head
    Neck
    Greater trochanter
    Lesser trochanter
    Medial condyle
    Lateral condyle
    Patellar surface
    Patella

Tibia
    Medial condyle
    Lateral condyle
        Tibial tuberosity
        Medial malleolus
Fibula
    Head
    Lateral malleolus
Tarsals
    Talus
    Calcaneus
Metatarsals
Phalanges
Joints
    Sacroiliac
    Hip
    Knee
    Tibiofibular (proximal and distal)
    Ankle (talocrural joint)
Muscles
Deep Hip Rotators
Gluteal Muscles
    Gluteus medius
    Gluteus maximus
Tensor Fasciae Latae
Hip Flexors
    Iliopsoas (iliacus and psoas major)

Medial Thigh Muscles (adductors)
    Pectineus
    Adductors (longus, brevis,
        magnus, minimus)
    Gracilis
Anterior Thigh Muscles
    Quadriceps femoris (rectus femoris,
        vastus medialis, lateralis, and
        intermedius)
    Sartorius
Posterior Thigh Muscles
    Biceps femoris
    Semimembranosus
    Semitendinosus
Anterior Leg Muscles
    Tibialis anterior
    Extensor digitorum longus
    Extensor hallucis longus
Lateral Leg Muscles (everters)
    Fibularis (longus and brevis)
Posterior Leg Muscles
    Tibialis posterior
    Flexor digitorum longus
    Flexor hallucis longus
    Popliteus
    Soleus
    Gastrocnemius
    Plantaris

Pelvic Girdle

The pelvic girdle is composed of two halves, each consisting of three bones—the ilium, ischium, and pubis. The three bones fuse to form a single bone, the os coxa or hip bone. Use the illustration below to identify the boundaries of the three bones. Each os coxa forms a strong articulation with the sacral vertebral column and the two bones articulate ventrally at the symphysis pubis. The sturdy os coxae form a deep joint with the femur. Compare the socket of the hip joint with the socket of the shoulder joint.

Os coxa or Hip Bone

os = bone, coxa = hip

This bone consists of the fused ilium, ischium, and pubis.

FIGURE 5.1 Right os coxa, lateral view

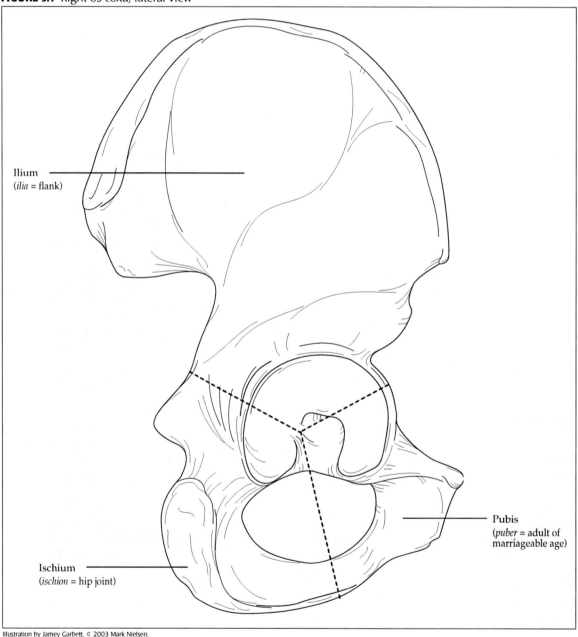

Ilium
(*ilia* = flank)

Pubis
(*puber* = adult of marriageable age)

Ischium
(*ischion* = hip joint)

Illustration by Jamey Garbett. © 2003 Mark Nielsen.

Os coxa

FIGURE 5.2 Os coxae, anterior view

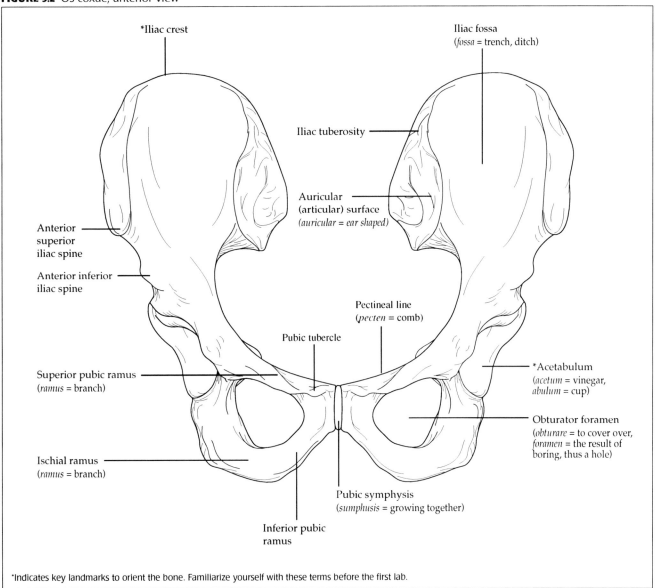

*Iliac crest

Iliac fossa
(*fossa* = trench, ditch)

Iliac tuberosity

Auricular
(articular) surface
(*auricular* = ear shaped*)

Anterior
superior
iliac spine

Anterior inferior
iliac spine

Pectineal line
(*pecten* = comb)

Pubic tubercle

*Acetabulum
(*acetum* = vinegar,
abulum = cup)

Superior pubic ramus
(*ramus* = branch)

Obturator foramen
(*obturare* = to cover over,
foramen = the result of
boring, thus a hole)

Ischial ramus
(*ramus* = branch)

Pubic symphysis
(*sumphusis* = growing together)

Inferior pubic
ramus

*Indicates key landmarks to orient the bone. Familiarize yourself with these terms before the first lab.

Illustration by Jamey Garbett. © 2003 Mark Nielsen.

Os coxa

FIGURE 5.3 Right os coxa, lateral view

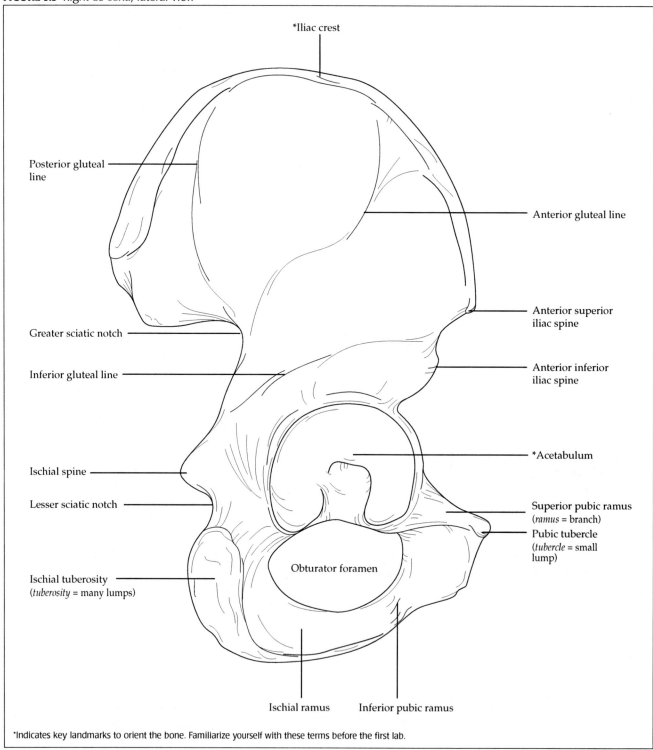

*Iliac crest

Posterior gluteal line

Anterior gluteal line

Greater sciatic notch

Anterior superior iliac spine

Inferior gluteal line

Anterior inferior iliac spine

Ischial spine

*Acetabulum

Lesser sciatic notch

Superior pubic ramus (*ramus* = branch)

Pubic tubercle (*tubercle* = small lump)

Ischial tuberosity (*tuberosity* = many lumps)

Obturator foramen

Ischial ramus

Inferior pubic ramus

*Indicates key landmarks to orient the bone. Familiarize yourself with these terms before the first lab.

Illustration by Jamey Garbett. © 2003 Mark Nielsen.

The Lower Limb Proper

The inferior limb proper consists of three basic regions—the thigh, crus (leg), and pes (foot). There are a total of 31 bones in each lower limb (pelvic girdle and inferior limb proper combined).

Femur

fero = to bear

The bone of the thigh. This is the largest bone in the body.

Patella

patina = small, shallow dish

The knee cap bone. The patella is a sesamoid bone. Sesamoid bones are bones that form within a tendon. The patella forms in the tendon of the quadriceps femoris musculature.

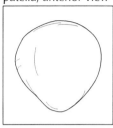

FIGURE 5.4 Right patella, anterior view

FIGURE 5.5 Right femur, anterior view (left), Right femur, posterior view (right)

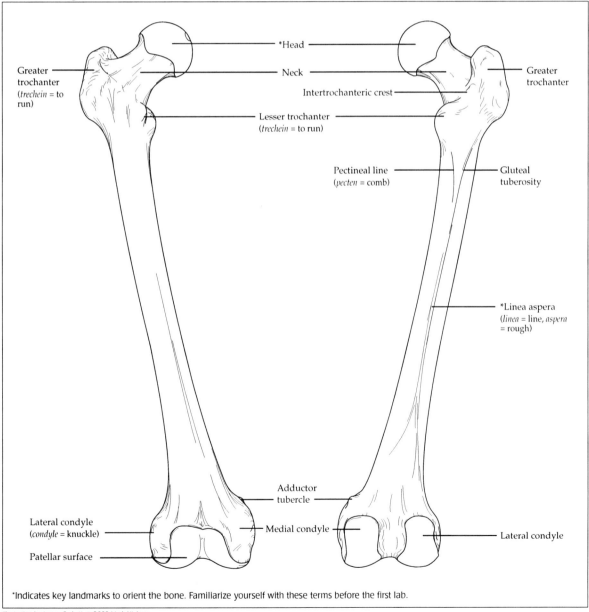

*Indicates key landmarks to orient the bone. Familiarize yourself with these terms before the first lab.

Illustration by Jamey Garbett. © 2003 Mark Nielsen.

Tibia

tibia = pipe or flute

Medial bone of the leg.

Fibula

figo = to fasten; came to mean buckle, brooch, or clasp

Lateral bone of the leg.

FIGURE 5.6 Anterior view: right fibula, right tibia (left), Posterior view: right fibula, right tibia (right)

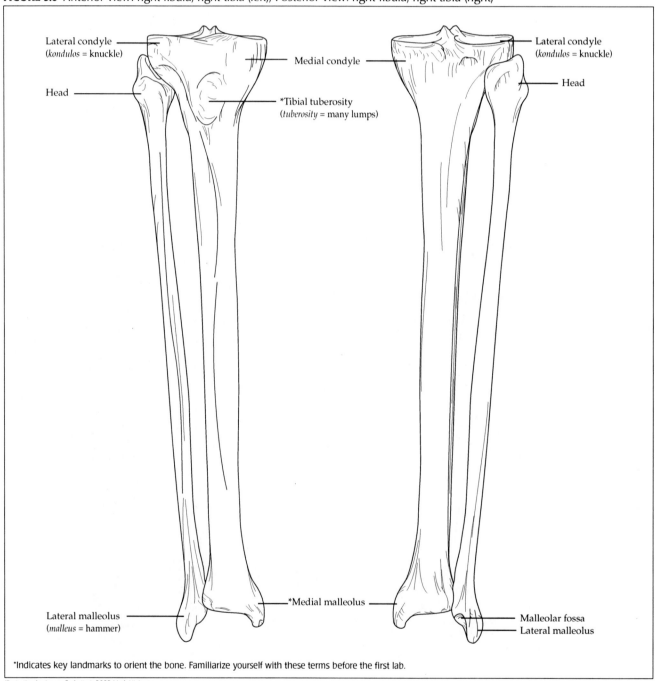

Lateral condyle
(*kondulos* = knuckle)

Head

Medial condyle

*Tibial tuberosity
(*tuberosity* = many lumps)

Lateral condyle
(*kondulos* = knuckle)

Head

*Medial malleolus

Lateral malleolus
(*malleus* = hammer)

Malleolar fossa
Lateral malleolus

*Indicates key landmarks to orient the bone. Familiarize yourself with these terms before the first lab.

Foot or Pes

The foot skeleton consists of three regions: the tarsus or ankle, the metatarsal region, and the digits or toes.

Tarsal Bones

tarsus = ankle

The seven bones in the proximal end of the foot skeleton that form the region known as the ankle.

Metatarsal Bones

meta = beyond, next to; tarsus = ankle

The five short long bones that form the middle portion of the foot skeleton.

Phalanges

phalanx = soldiers lined up for battle

The fourteen bones of the digits (toes). Each small toe has three phalanges, while the great toe has only two.

FIGURE 5.7

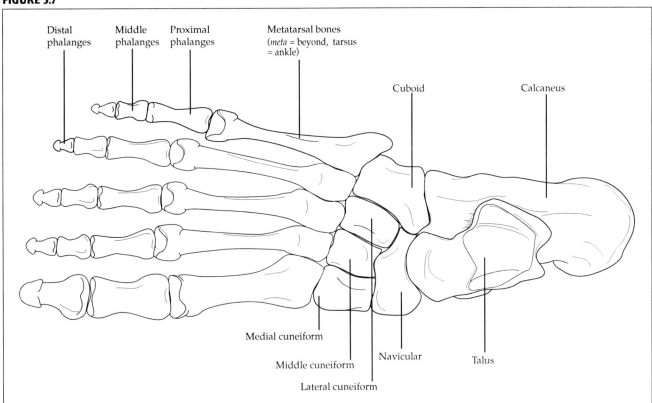

Illustration by Jamey Garbett. © 2003 Mark Nielsen.

Sexual Dimorphism

Skeletal sexual dimorphism refers to the diagnostic differences that exist between the adult male and female skeleton. There are many distinguishing points of difference that exist between male and female skeletal specimens. Some of the most prominent differences can be demonstrated within the bones of the pelvis. The morphological differences that are evidenced here reflect functional differences between males and females. A wide birth canal for the reproductive function of the female and the more muscular nature of the male are manifested in the anatomy of the pelvis. Compare the illustrations below and the pelvic models in the lab. Notice the wider pelvic opening in the female. As a result of the wider pelvic opening, the angle formed between the inferior rami of the adjoining pubic bones tends to be obtuse (greater than 90 degrees) in the female while being acute (less than 90 degrees) in the male. The male pelvis tends to have more prominent sites for muscle attachment, reflecting the more extensive muscle development of the male. What other differences do you notice?

FIGURE 5.8 Male pelvis, superior view to left, lateral view to right

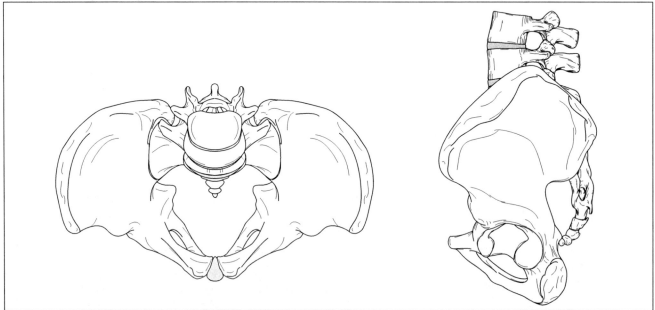

Illustration by Jamey Garbett. © 2003 Mark Nielsen.

FIGURE 5.9 Female pelvis, superior view to left, lateral view to right

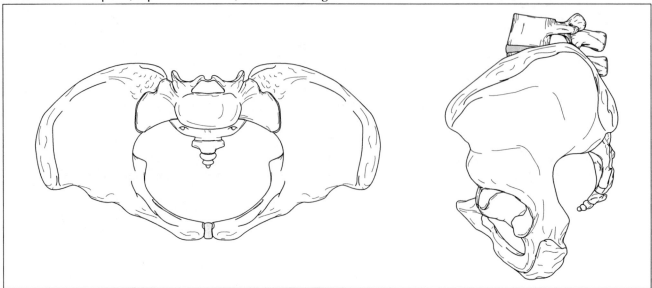

Illustration by Jamey Garbett. © 2003 Mark Nielsen.

Hip Joint
Movements of the Hip
Flexion

Extension

Abduction

Adduction

Medial rotation

Lateral rotation

FIGURE 5.10

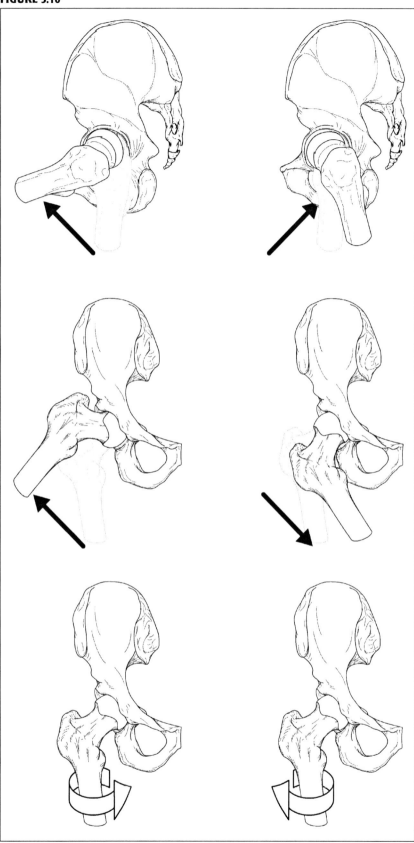

Summary of the Deep Hip Rotator Muscles

Common attachments

Common actions at joints

Practical applications

Illustration by Jamey Garbett. © 2003 Mark Nielsen.

FIGURE 5.11

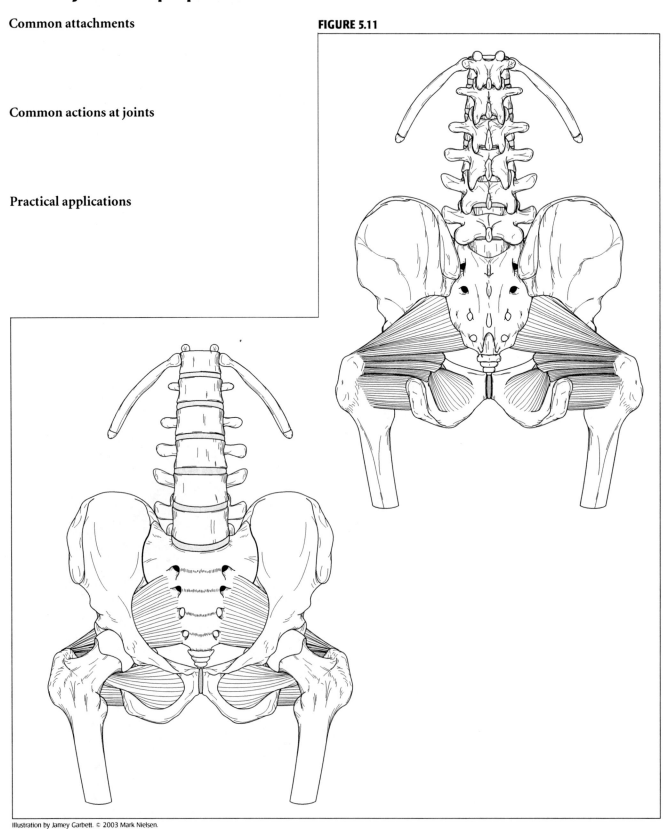

Deep Hip Rotator Muscles

Muscle	Origin	Insertion	Function	Innervation
Piriformis	Anterior surface of the sacrum between the anterior sacral foramina; external surface of posterior iliac spine and the capsule of the sacro-iliac joint	Medial side of the superior border of greater trochanter of the femur	Lateral rotation of the extended thigh; abducts the flexed thigh	Nerve to piriformis (S1 and S2), some-times contributions from L5
Superior gemellus	Dorsal surface of the spine of the ischium	Joins the upper edge of the obturator internus tendon to the medial surface of the greater trochanter of the femur	Lateral rotation of the hip joint; helps abduct a flexed thigh	Nerve to obturator internus (L5 and S1)
Obturator internus	Internal surface of the obturator membrane and the surrounding bone from the pelvic brim to the base of the pubo-ischial ramus and the pubic body to the edge of the greater sciatic notch	Anteriorly on the medial surface of greater trochanter of femur above the trochanteric fossa	Lateral rotation of the hip joint; helps abduct a flexed thigh	Nerve to obturator internus (L5 and S1)
Inferior gemellus	Upper portion of the tuberosity of the ischium	Joins the inferior edge of the obturator internus tendon to the medial surface of the greater trochanter of the femur	Lateral rotation of the hip joint; helps abduct a flexed thigh	Nerve to quadratus femoris (L5 and S1)
Quadratus femoris	Superior portion of the lateral surface of the ischial tuberosity	Tubercle above the middle part of the trochanteric crest on the posterior side of the femur	Lateral rotation of the hip	Nerve to quadratus femoris (L5 and S1)
Obturator externus	Medial two-thirds of the external surface of the obturator membrane and adjacent surfaces of the pubic body and pubic and ischial rami	Trochanteric fossa on the medial aspect of the greater trochanter of the femur	Lateral rotation of the hip joint; helps abduct a flexed thigh	Obturator nerve (L3 and L4)

Gluteal Muscles

Gluteus Minimis Muscle

Gloutos = rump, minimis = the smallest

Origin

Insertion

Function

Structure and
relationships

FIGURE 5.12

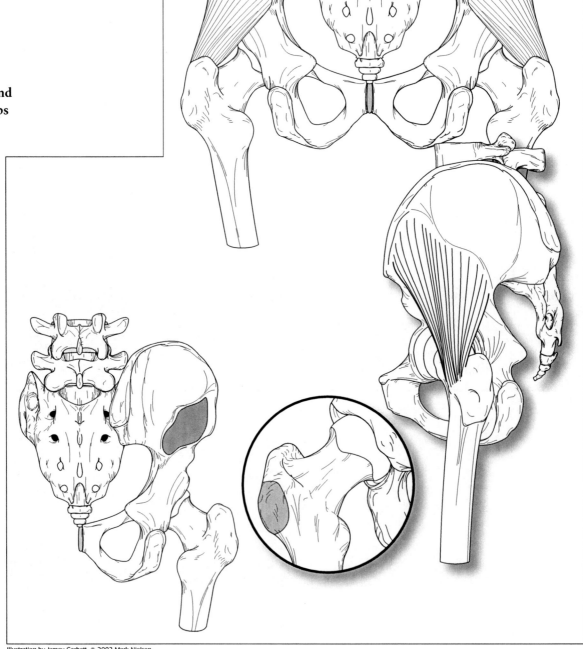

Illustration by Jamey Garbett. © 2003 Mark Nielsen.

Gluteus Medius Muscle

Gloutos = rump, medius = the middle one

Origin

Insertion

Function

Structure and relationships

FIGURE 5.13

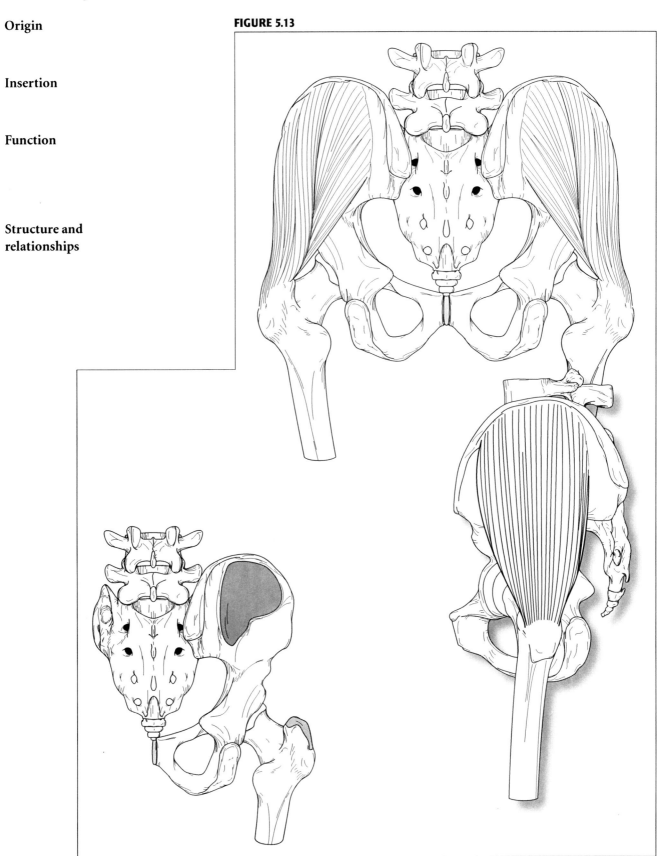

Illustration by Jamey Garbett. © 2003 Mark Nielsen.

Gluteus Maximus Muscle

Gloutos = rump, maximus = the largest

Origin

Insertion

Function

Structure and relationships

FIGURE 5.14

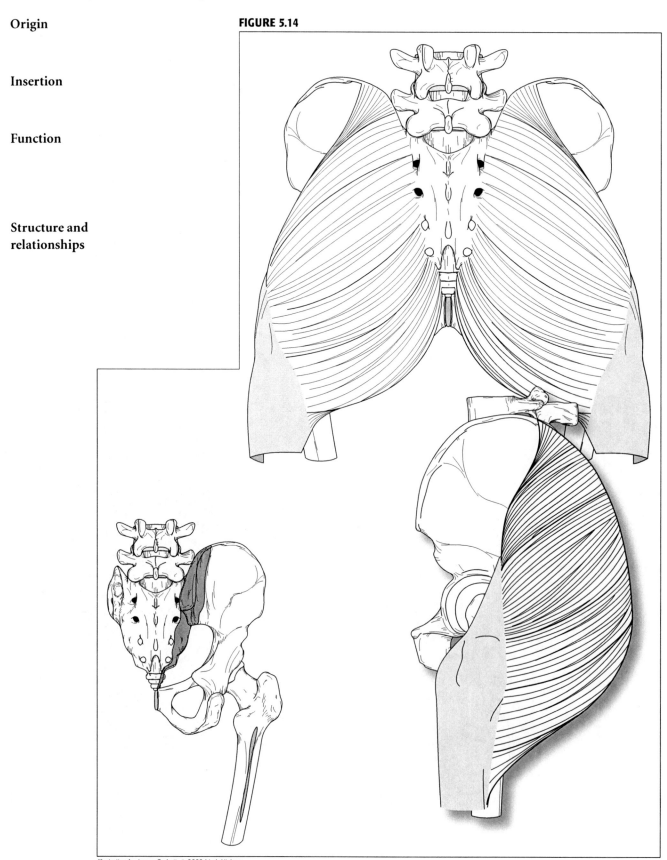

Illustration by Jamey Garbett. © 2003 Mark Nielsen.

Tensor Fasciae Latae Muscle

tensor = to tense, *fasciae latae* = referring to the fasciae
latae of the thigh

Origin

Insertion

Function

Structure and
relationships

FIGURE 5.15

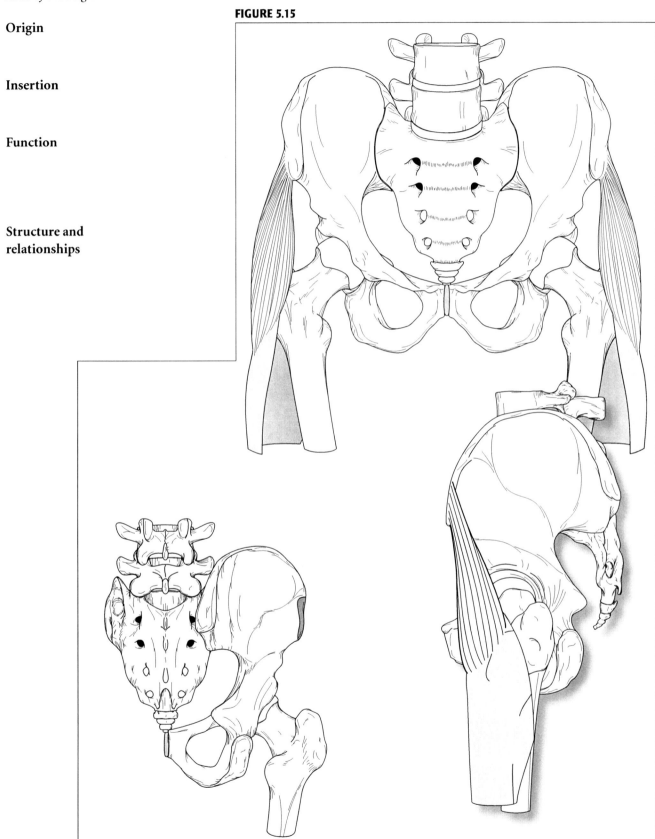

Illustration by Jamey Garbett. © 2003 Mark Nielsen.

Summary of the Gluteal Muscles

Common attachments

FIGURE 5.16

Common actions at joints

Practical applications

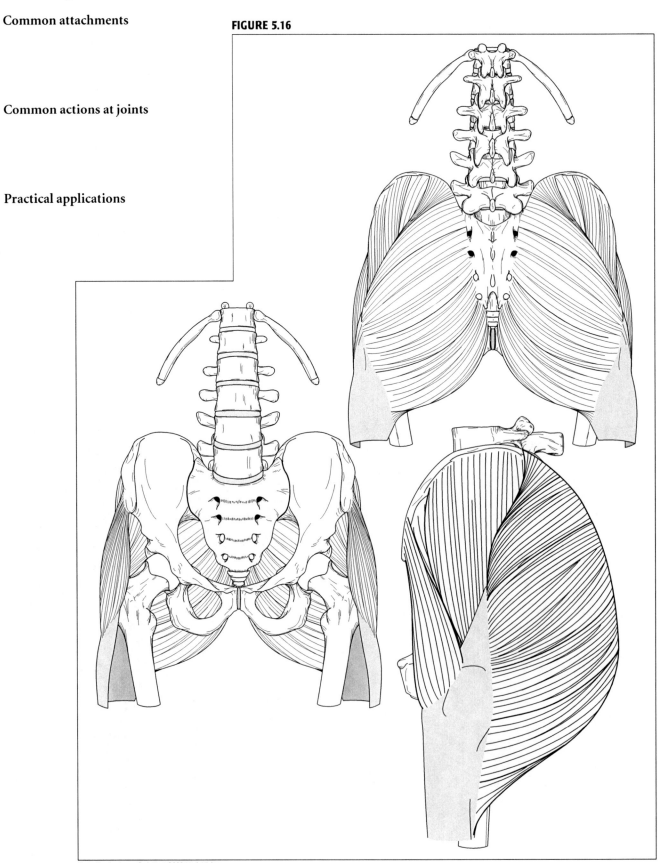

Illustration by Jamey Garbett. © 2003 Mark Nielsen.

Gluteal Muscles

Muscle	Origin	Insertion	Function	Innervation
Gluteus minimis	Between the anterior and inferior gluteal lines of the ilium	Ridge on the anterior and lateral surface of the greater trochanter of the femur; capsule of the hip joint	Abductor of the hip joint (important in supporting the upright hip when opposite foot is raised off of the ground); anterior fibers rotate the hip joint medially	Superior gluteal nerve (L4, L5, and S1)
Gluteus medius	Between the anterior and posterior gluteal lines and the overlying gluteal fascia	Ridge on the lateral surface of the greater trochanter of the femur	Powerful abductor of the hip joint (important in supporting the upright hip when opposite foot is raised off of the ground); anterior fibers rotate the hip joint medially	Superior gluteal nerve (L4, L5, and S1)
Gluteus maximus	From the rough ilium behind the posterior gluteal line, the posterior iliac crest, the lower posterior sacrum, and the side of the coccyx, and the fascia of the gluteus medius	Large upper two-thirds of the muscle attaches into the iliotibial tract of the fascia lata; lower third to the gluteal tuberosity	Extends the flexed thigh and lateral rotation of hip (can move the trunk or thigh, depending on which is fixed); Upper fibers assist with abduction of the hip joint; it tenses the iliotibial tract and the fascia lata and accordingly stabilizes the femur on the tibia	Inferior gluteal nerve (L5, S1, and S2)
Tensor fasciae latae	Outer lip of the anterior iliac crest; lateral aspect of the anterior superior iliac spine; underside of the iliotibial tract	Descends between the two layers of the iliotibial tract and attaches to them (typically ends in the upper third of the thigh, but can extend to the level of the femoral condyles)	Medial rotation and flexion of the hip; assists weakly with extension of the knee; lateral rotation of the leg at the knee; stabilizes the hip and the knee during standing	Superior gluteal nerve (L4, L5, and S1)

Hip Flexor Muscles

Iliopsoas Muscle

Psoas Major Muscle

psoa = loin muscle, major = greater

Origin

Insertion

Function

Structure and
relationships

FIGURE 5.17

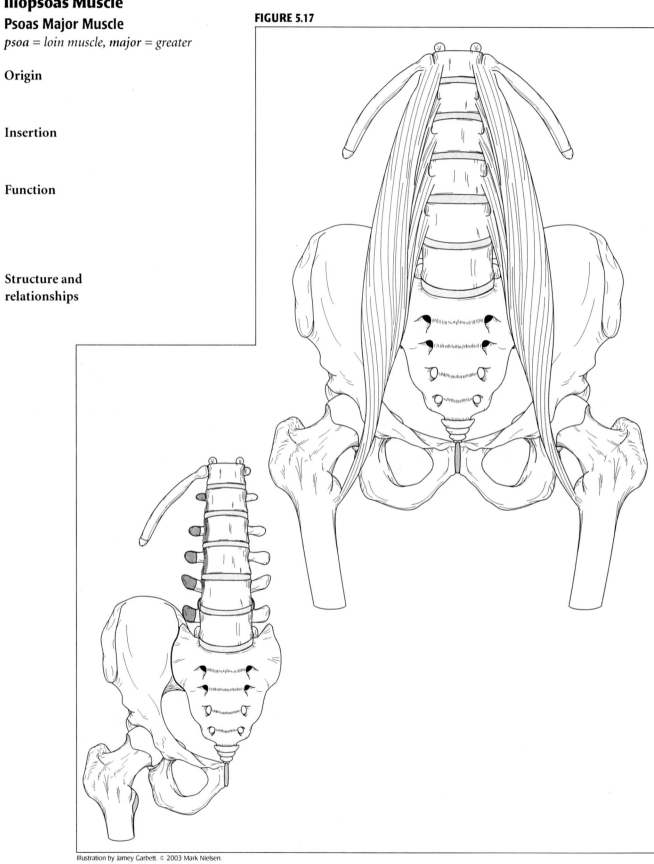

Illustration by Jamey Garbett. © 2003 Mark Nielsen.

Iliacus Muscle

iliac = referring to the ilium

Origin

Insertion

Function

**Structure and
relationships**

FIGURE 5.18

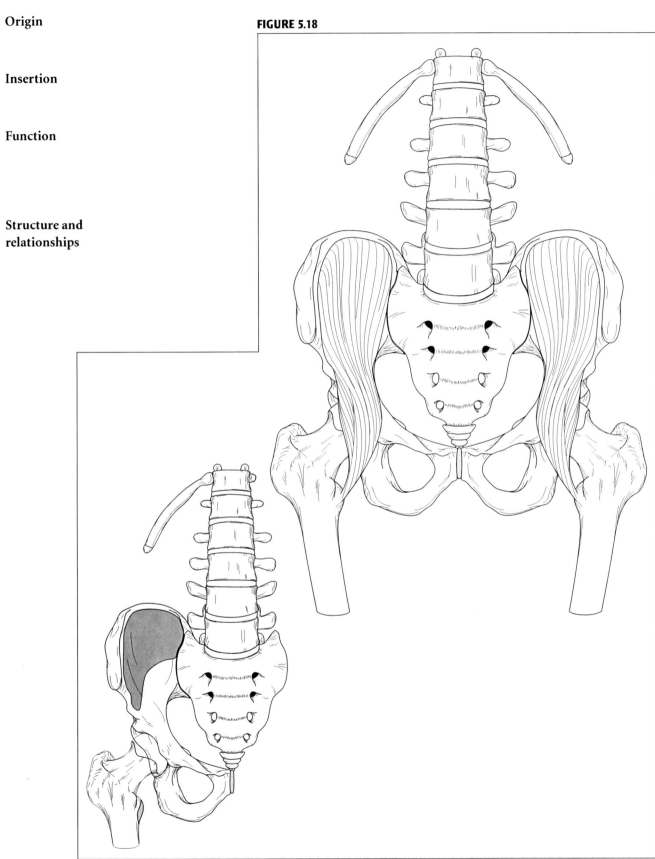

Illustration by Jamey Garbett. © 2003 Mark Nielsen.

Summary of the Hip Flexor Muscles

FIGURE 5.19

Common attachments

Common actions at joints

Practical applications

Thinking about pulleys

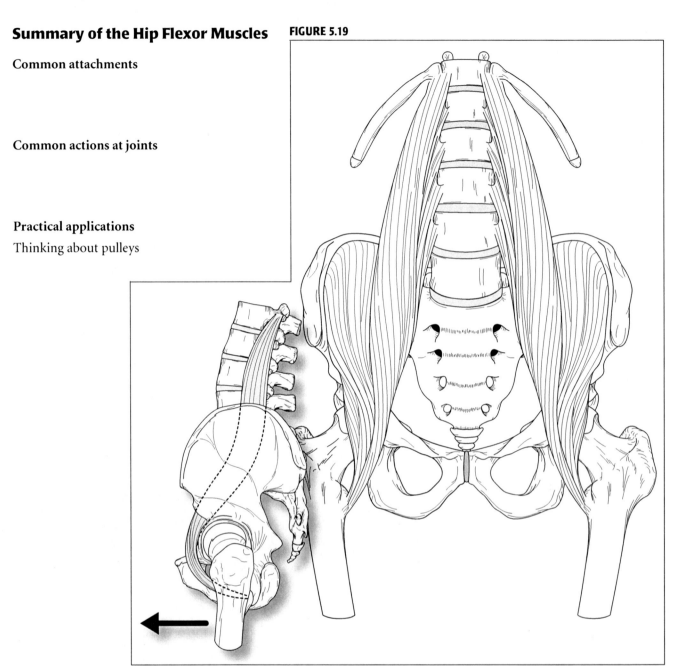

Illustration by Jamey Garbett. © 2003 Mark Nielsen.

Hip Flexor Muscles

Muscle	Origin	Insertion	Function	Innervation
Psoas major	Transverse processes of lumbar vertebrae and lateral aspect of lumbar vertebral bodies	Lesser trochanter of the femur	Powerful flexor of the hip; weak lateral rotator of the hip joint; Extends the lumbar vertebral column (deepening the lumbar curve) to maintain posture	Branches from lumbar spinal nerves 1, 2, and 3
Iliacus	Superior half of the iliac fossa, the inner lip of the iliac crest, the anterior ligaments of the sacroiliac joint, and the superior surface of the lateral portion of the sacrum	Converges with the psoas major tendon to the lesser trochanter of femur; some fibers pass below the trochanter onto the medial surface of the proximal femur	Powerful flexor of the hip; weak lateral rotator of the hip joint	Via branches from femoral nerve in pelvic region (L2 and L3)

Muscles of the Thigh

The muscles of the thigh form a large mass of real estate. These muscles occur within three distinct muscle compartments and produce joint movements at both the hip and the knee joints.

Medial Thigh Compartment Muscles

Pectineus Muscle

pecten = comb

Origin

Insertion

Function

Structure and relationships

FIGURE 5.20

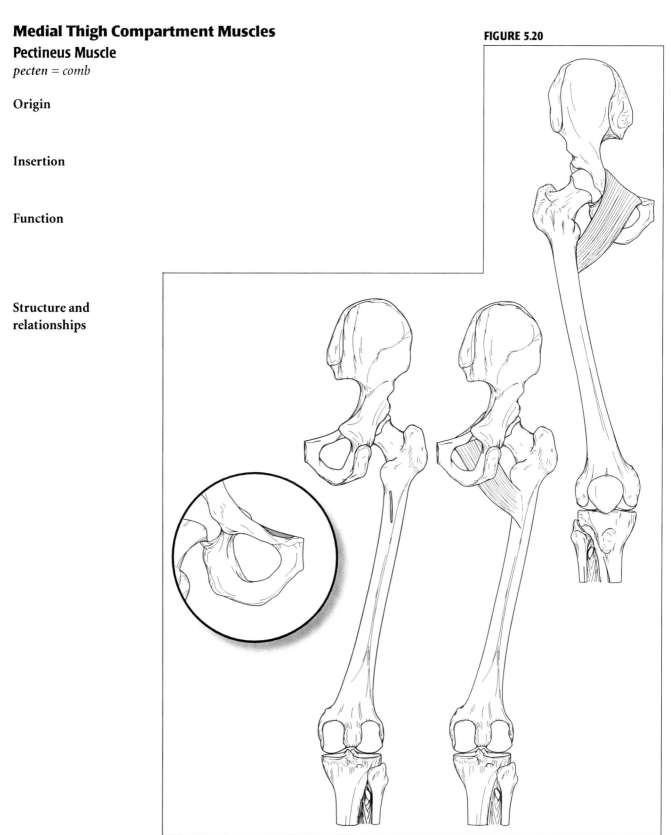

Illustration by Jamey Garbett. © 2003 Mark Nielsen.

Adductor Brevis Muscle
adductor = one that adducts, *brevis* = short

Origin

Insertion

Function

Structure and relationships

FIGURE 5.21

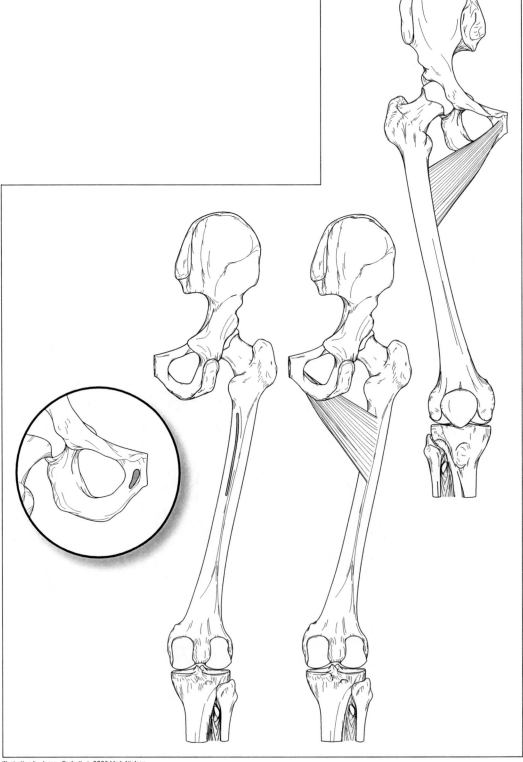

Illustration by Jamey Garbett. © 2003 Mark Nielsen.

Adductor Longus Muscle

adductor = one that adducts, **longus** *= long*

Origin

Insertion

Function

Structure and relationships

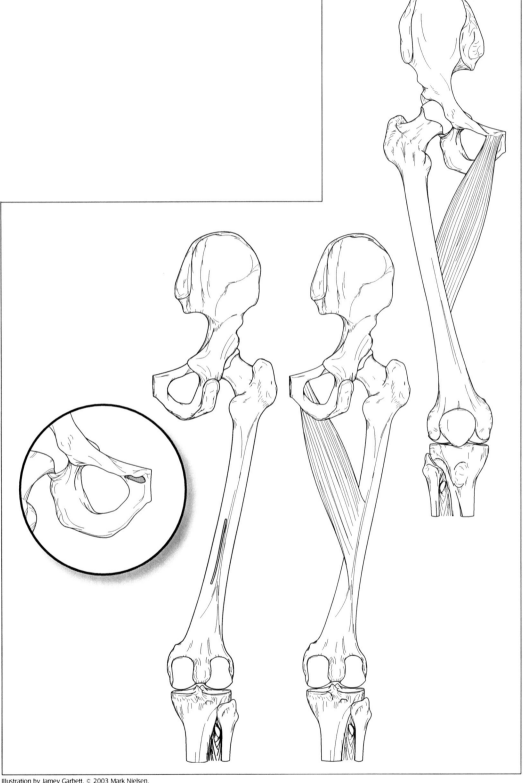

FIGURE 5.22

Illustration by Jamey Garbett. © 2003 Mark Nielsen.

Adductor Minimis Muscle

adductor = one that adducts, minimis = the smallest

Origin

Insertion

Function

Structure and relationships

FIGURE 5.23

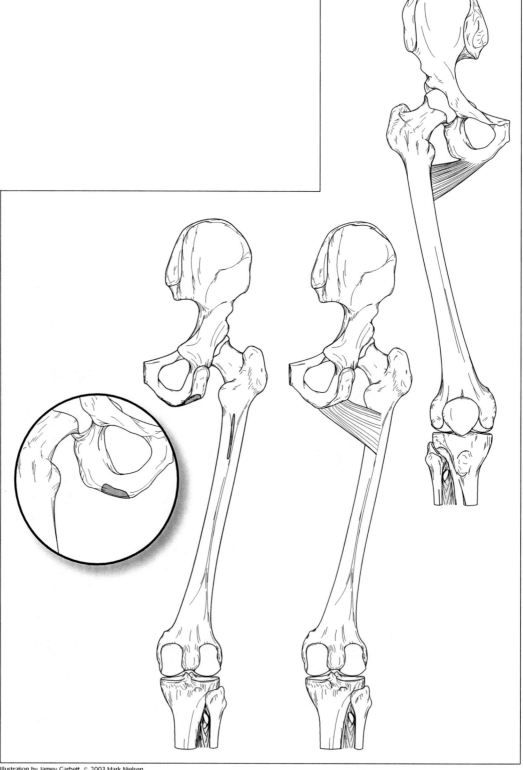

Illustration by Jamey Garbett. © 2003 Mark Nielsen.

Adductor Magnus Muscle
adductor = one that adducts, ***magnus*** = the largest

Origin

Insertion

Function

Structure and relationships

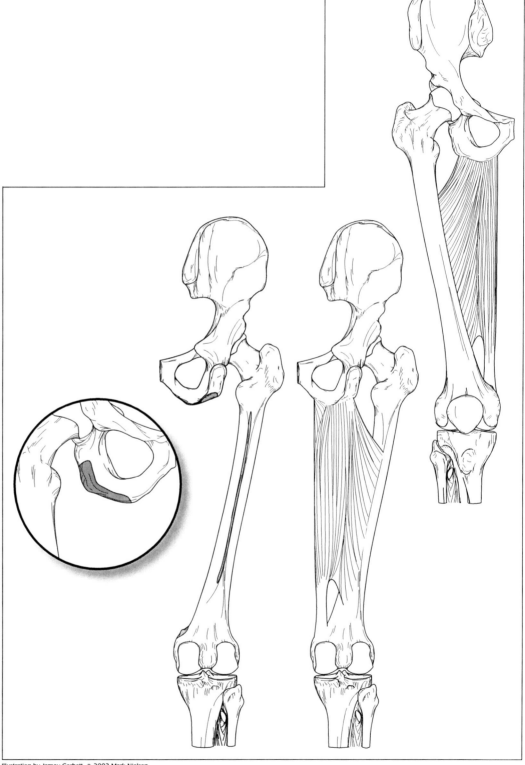

FIGURE 5.24

Illustration by Jamey Garbett. © 2003 Mark Nielsen.

Gracilis Muscle

gracilis = thin, slender

Origin

Insertion

Function

Structure and relationships

FIGURE 5.25

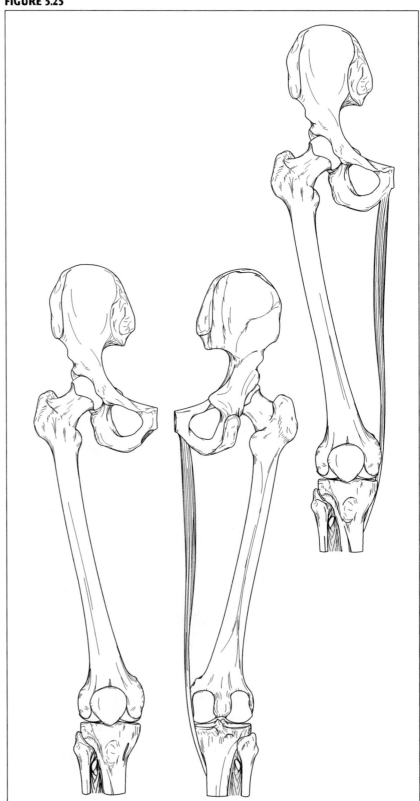

Illustration by Jamey Garbett. © 2003 Mark Nielsen.

Summary of the Medial Thigh Compartment Muscles

Common attachments

Common actions at joints

FIGURE 5.26

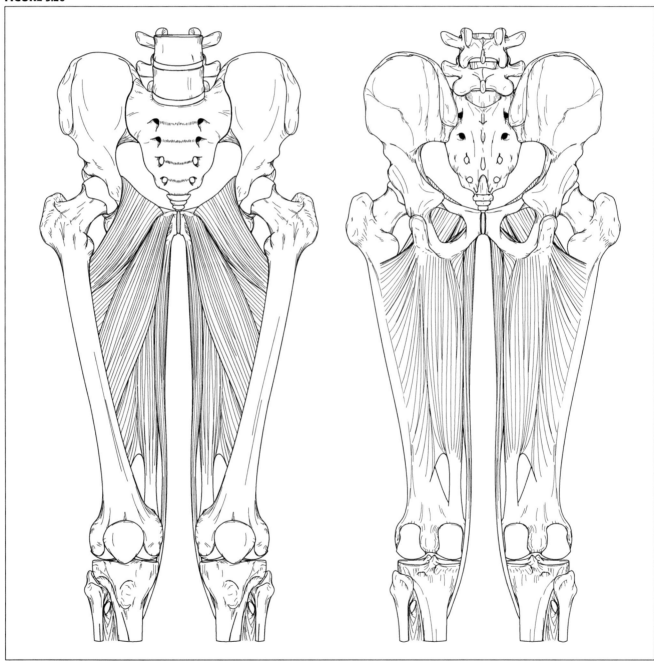

Illustration by Jamey Garbett. © 2003 Mark Nielsen.

Muscles of the Medial Compartment of the Thigh

Muscle	Origin	Insertion	Function	Innervation
Pectineus	Pectineal line of the pubis and the bone just anterior to this line	Pectineal line of the femur	Adduction and flexion of the hip joint	Femoral nerve (L2 and L3); Accessory obturator nerve (L3) (variable, not always present)
Adductor brevis	Narrow strip of the external surface of the pubic body and inferior ramus	Pectineal line and proximal part of the linea aspera of the femur	Adduction and lateral rotation of hip	Obturator nerve (L2 and L3)
Adductor longus	Anterior surface of the pubis between the crest and the symphysis	Middle third of the linea aspera of the femur	Adduction, flexion, and lateral rotation of hip	Obturator nerve (L2 and L3)
Adductor minimis	External surface of the inferior pubic ramus	Medial margin of the gluteal tuberosity of the femur	Adduction and lateral rotation of the hip	Obturator nerve (L2)
Adductor magnus	External and inferior surface of the ischial ramus of the pubis and lateral surface of the ischial tuberosity	The length of the linea aspera and the proximal part of the medial supracondylar line; the adductor tubercle of the femur	Adduction and extension of hip; upper fibers laterally rotate	Obturator nerve (L2 and L3) and tibial nerve (L4)
Gracilis	Lower half of the body of the pubis; inferior ramus of the pubis and a small adjacent portion of the ischial ramus	Anteromedial surface of the proximal tibia just below the medial condyle; fascia of the leg; attaches slightly proximal to the semitendinosus and just behind the sartorius	Flexion of the knee and rotates the knee medially; assists with adduction of the hip	Obturator nerve (L2 and L3)

Practical applications

Functional Aspects of the Muscles of the Hip Joint
Gluteal Muscle Actions

Gluteus medius and minimis

Gluteus maximus

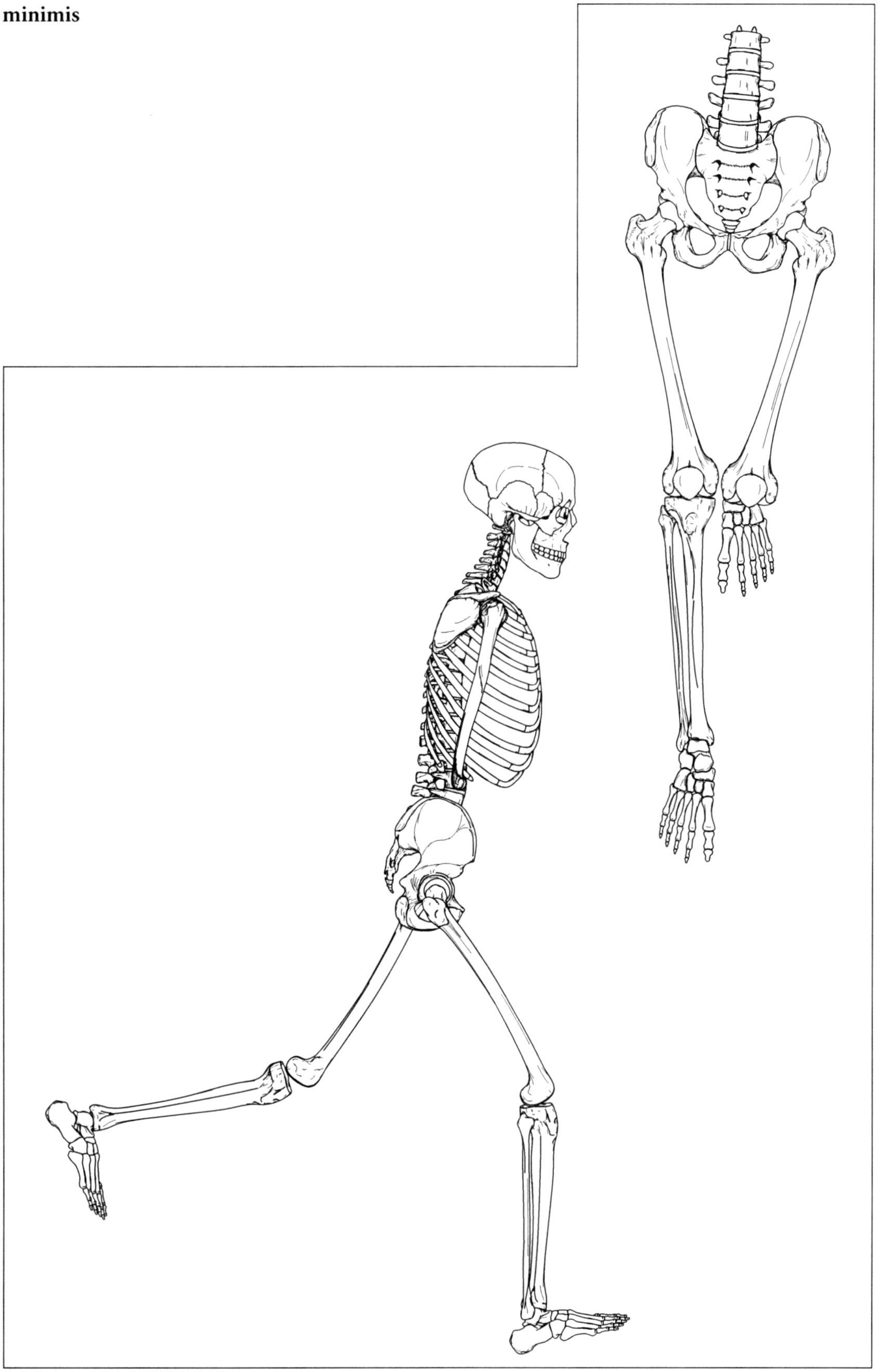

Illustration by Jamey Garbett. © 2003 Mark Nielsen.

FIGURE 5.27

Why the predominance of lateral rotators of the hip?

FIGURE 5.28

Iliotibial Band

Structure and function

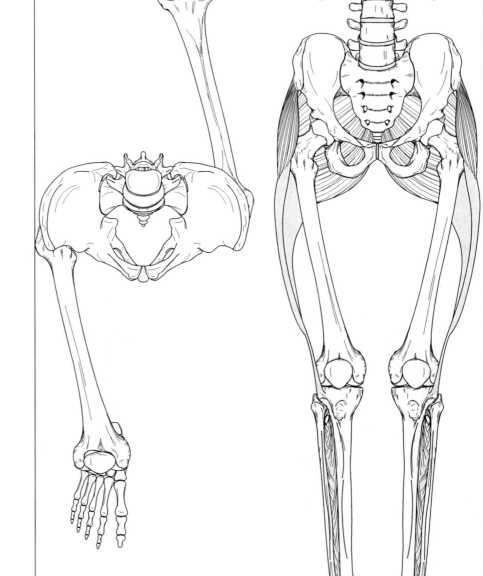

Illustration by Jamey Garbett. © 2003 Mark Nielsen.

Knee Joint

Before we explore the final two compartments of the thigh it is imperative to study the structure and function of the knee joint because the remaining two muscle compartments of the thigh are prime movers of the knee joint.

Type of Joint

Articulations

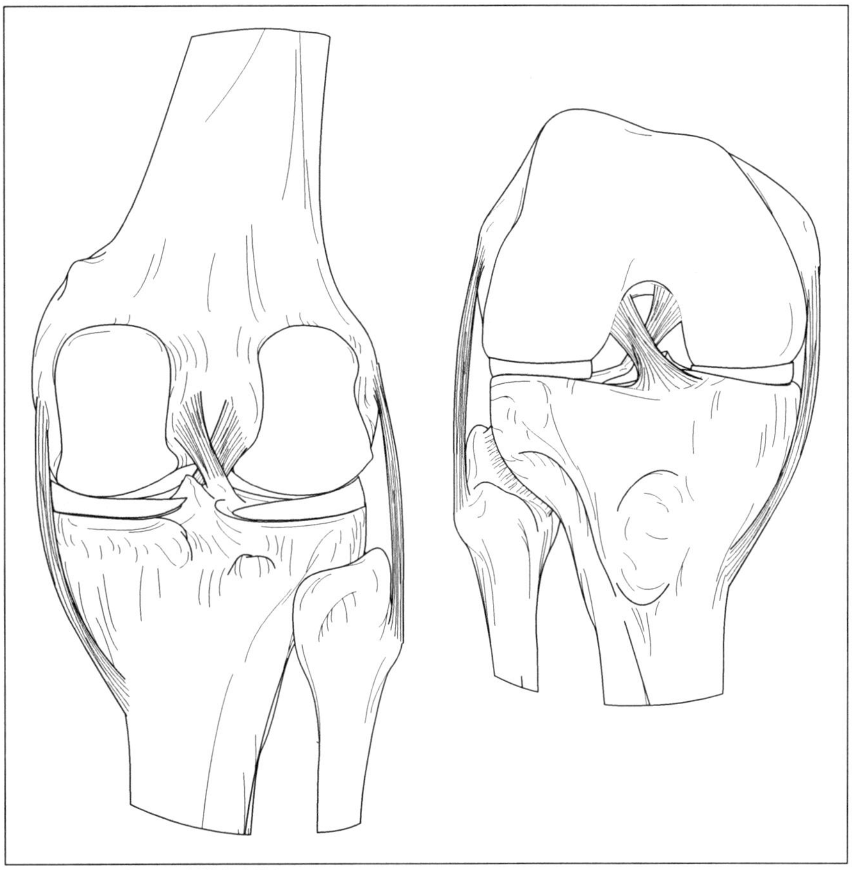

Ligaments

Fibular (lateral) collateral ligament

Tibial (medial) collateral ligament

Anterior cruciate ligament

Posterior cruciate ligament

Menisci

Medial meniscus

Lateral meniscus

Movements at the Knee Joint

Flexion

Extension

Rotation

FIGURE 5.29

Illustration by Jamey Garbett. © 2003 Mark Nielsen.

FIGURE 5.30

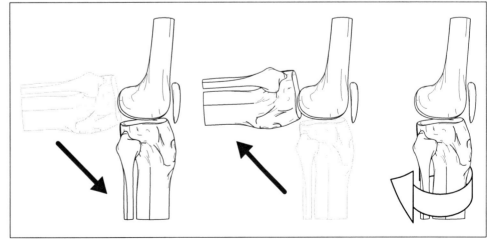

Illustration by Jamey Garbett. © 2003 Mark Nielsen.

Muscles of the Thigh

Anterior Thigh Compartment Muscles

Quadriceps Femoris Muscle

quad = four + kephale = head, femoris = referring to the femur

Vastus Intermedius Muscle

vastus = large, intermedius = intermediate

Origin

Insertion

Function

Structure and
relationships

FIGURE 5.31

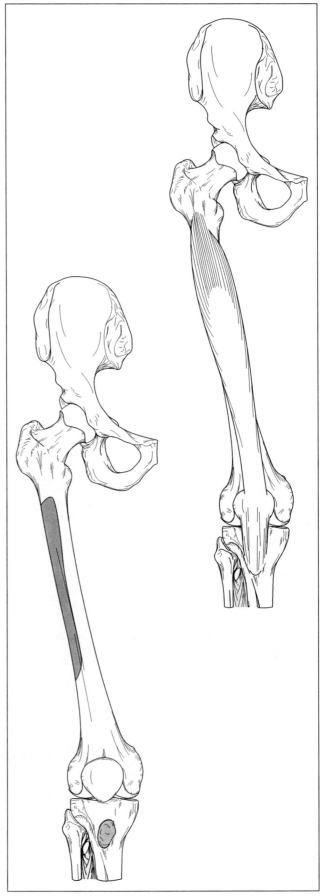

Illustration by Jamey Garbett. © 2003 Mark Nielsen.

Vastus Lateralis Muscle

vastus = large, lateralis = lateral

Origin

Insertion

Function

Structure and relationships

FIGURE 5.32

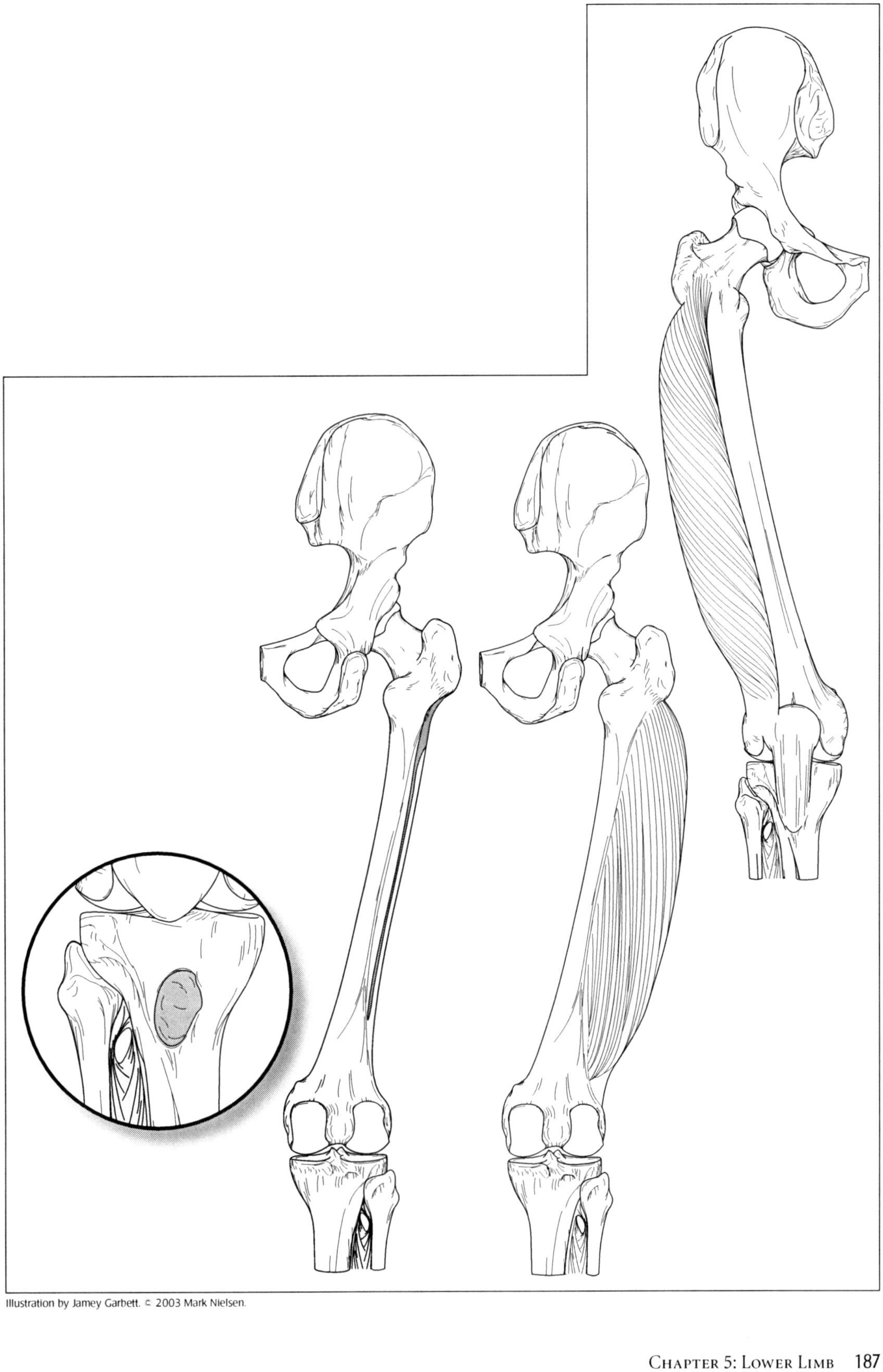

Illustration by Jamey Garbett. © 2003 Mark Nielsen.

Vastus Medialis Muscle

vastus = large, medialis = medial

Origin

Insertion

Function

Structure and relationships

FIGURE 5.33

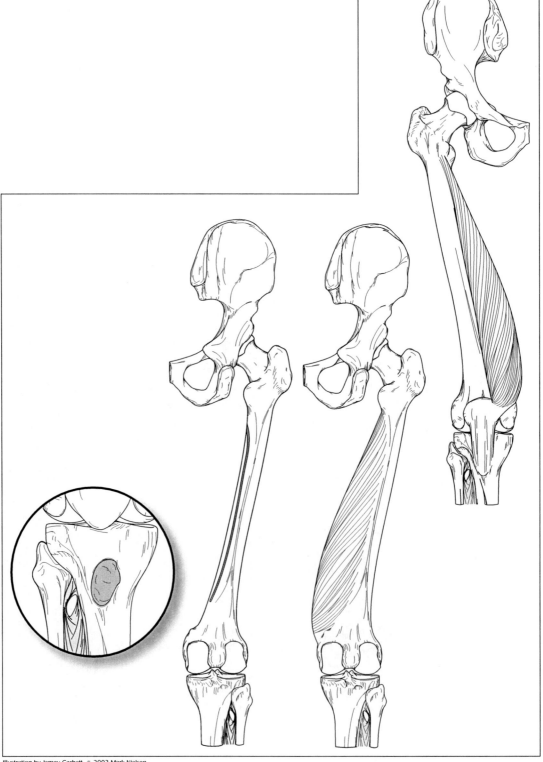

Illustration by Jamey Garbett. © 2003 Mark Nielsen.

Rectus Femoris Muscle

rectus = straight, femoris = referring to the femur

Origin

Insertion

Function

Structure and relationships

FIGURE 5.34

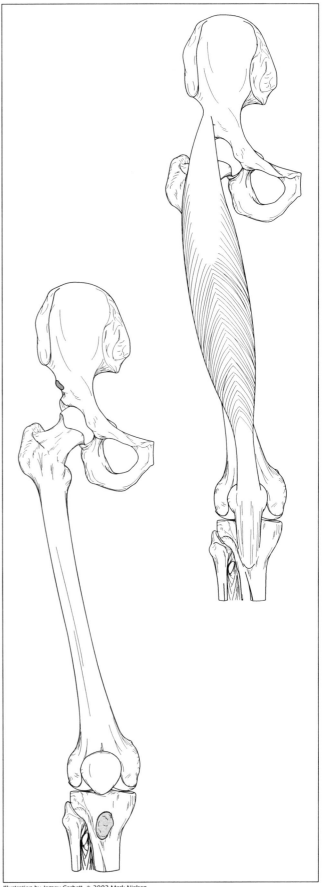

Illustration by Jamey Garbett. © 2003 Mark Nielsen.

Articularis Genus Muscle

articulo = *to join or connect,* ***genu*** = *referring to the knee*

Origin

Insertion

Function

Structure and relationships

FIGURE 5.35

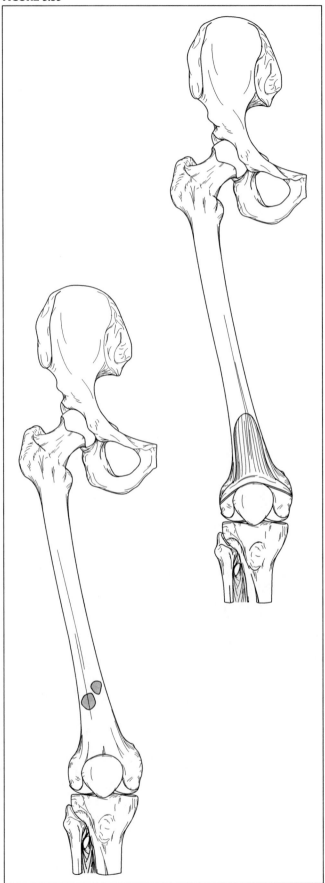

Illustration by Jamey Garbett. © 2003 Mark Nielsen.

Sartorius Muscle

sartor = tailor

Origin

Insertion

Function

Structure and relationships

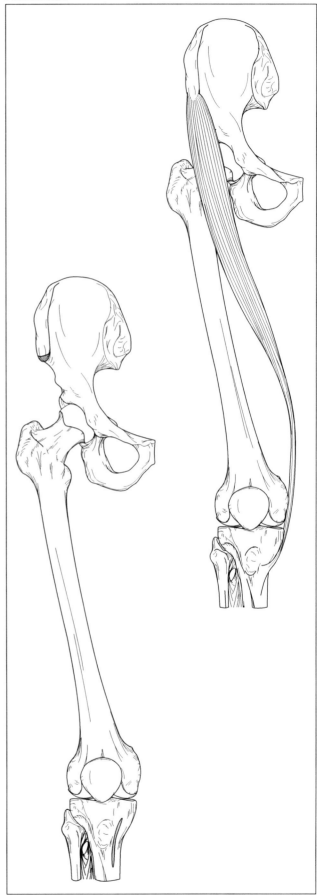

FIGURE 5.36

Illustration by Jamey Garbett. © 2003 Mark Nielsen.

Summary of the Anterior Thigh Compartment Muscles

Common attachments

Common actions at joints

FIGURE 5.37

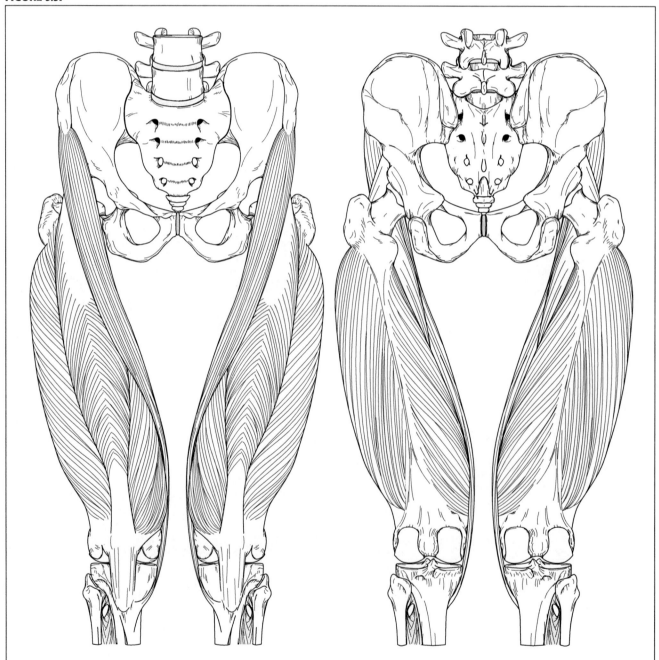

Illustration by Jamey Garbett. © 2003 Mark Nielsen.

Muscles of the Anterior Compartment of the Thigh

Muscle	Origin	Insertion	Function	Innervation
Vastus intermedius	Proximal two-thirds of anterior and lateral surfaces of the femoral shaft; lower part of the lateral intermuscular septum	Tibial tuberosity (see below); sends a tendinous expansion to the medial aspect of the joint capsule of the knee and the medial condyle of the tibia	Extension of the knee	Femoral nerve (L2, L3, and L4)
Vastus lateralis	Proximal portion of intertrochanteric line, anterior and inferior margins of the greater trochanter, lateral side of gluteal tuberosity and laterally on the linea aspera in its proximal half; fibers also arise from the lateral intermuscular septum	Tibial tuberosity (see below); sends a lateral slip to the capsule of the knee joint, the lateral condyle of the tibia, and the iliotibial tract	Extension of the knee	Femoral nerve (L2, L3, and L4)
Vastus medialis	Distal edge of the intertrochanteric line, the pectineal line of the femur, the medial side of entire linea aspera, the medial supracondylar line, and the medial intermuscular septum	Tibial tuberosity (see below); sends a tendinous expansion to the medial aspect of the joint capsule of the knee and the medial condyle of the tibia	Extension of the knee	Femoral nerve (L2, L3, and L4)
Rectus femoris	Anterior inferior spine of the ilium (straight tendon); superior groove of the acetabulum and from the anterior surface of the fibrous capsule of the hip joint (reflected tendon)	Tibial tuberosity (the patella is a sesamoid bone within the quadriceps tendon, the combined tendon of the rectus and vasti muscles; proximal to the patella the tendon is named the quadriceps tendon, distal to the patella the tendon is called the patellar ligament)	Flexion of the hip; extension of the knee	Femoral nerve (L2, L3, and L4)
Articularis genus	Distal end of anterior femoral shaft	Proximal extension of the joint capsule of the knee	Pulls the suprapatellar bursa of the knee joint capsule proximally to move it away from the patella during extension of the knee	Femoral nerve (L3 and L4)
Sartorius	Anterior superior iliac spine and the upper part of the notch below this landmark	Surface of the tibia medial to the tibial tuberosity and just below the medial condyle; sends a slip of tendon to the capsule of the knee joint; attaches just anterior to the gracilis and the semitendinosus	Flexion and lateral rotation of the hip; flexion of the knee	Femoral nerve (L2 and L3)

Practical applications

Posterior Thigh Compartment Muscles

Biceps Femoris Muscle

biceps = two heads, femoris = referring to the femur

Origin

Insertion

Function

Structure and
relationships

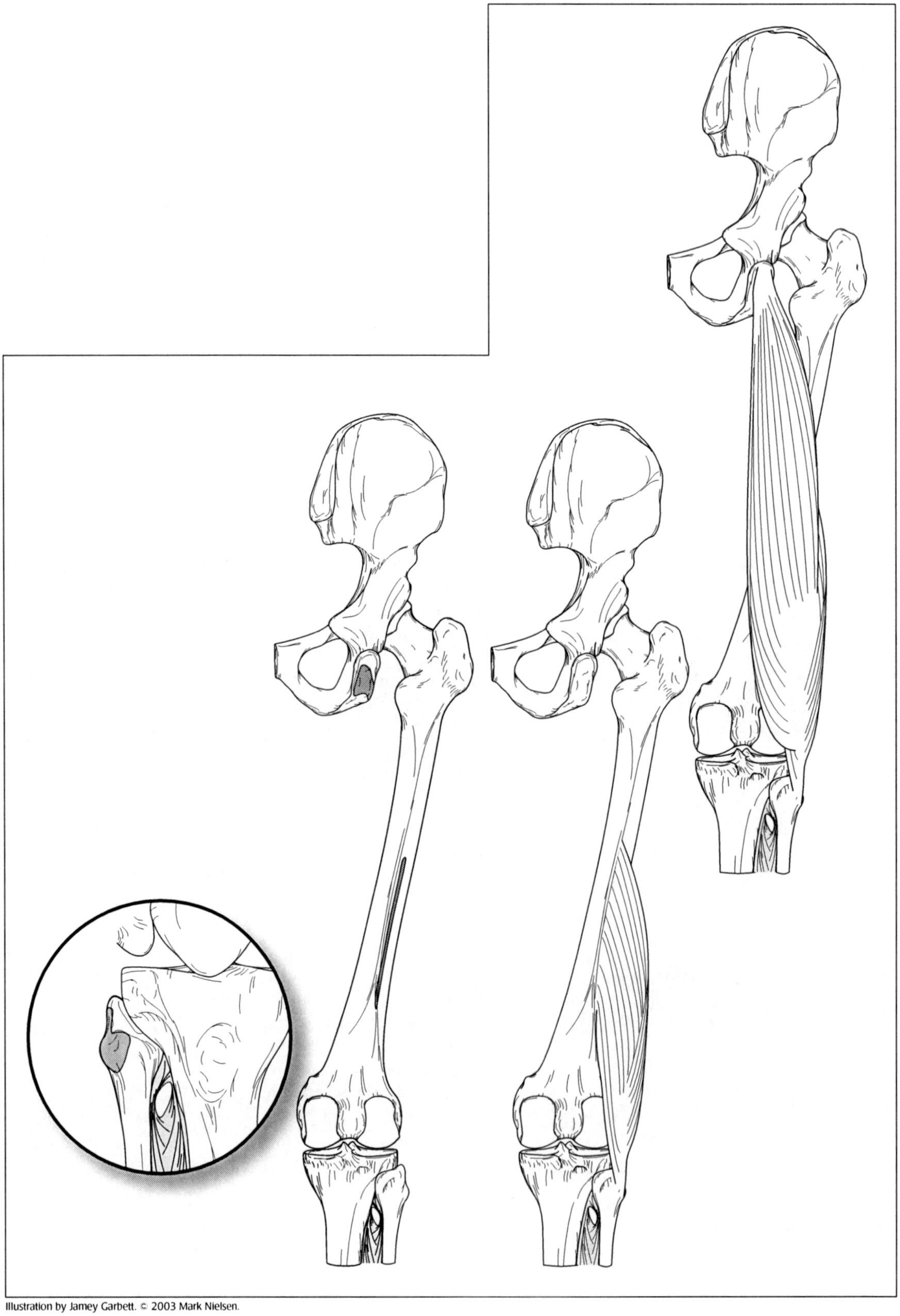

Illustration by Jamey Garbett. © 2003 Mark Nielsen.

FIGURE 5.38

Semimembranosus Muscle

semi- = *part* + ***membranosus*** = *membranous*

Origin

Insertion

Function

Structure and relationships

FIGURE 5.39

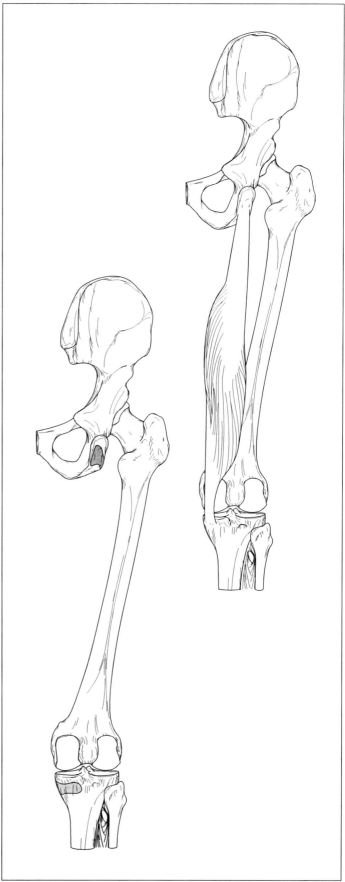

Illustration by Jamey Garbett. © 2003 Mark Nielsen.

Semitendinosus Muscle

semi- = *part* + *tendinosus* = *tendinous*

Origin

Insertion

Function

Structure and relationships

FIGURE 5.40

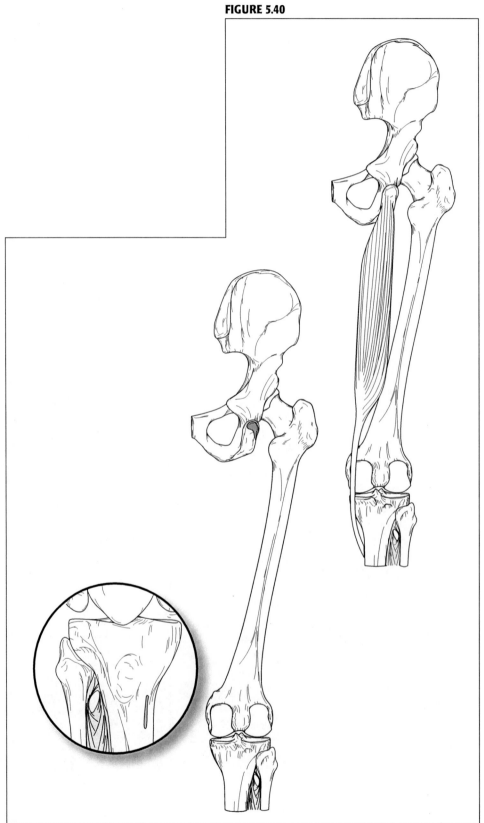

Illustration by Jamey Garbett. © 2003 Mark Nielsen.

Summary of the Posterior Thigh Compartment Muscles

Common attachments

Common actions at joints

FIGURE 5.41

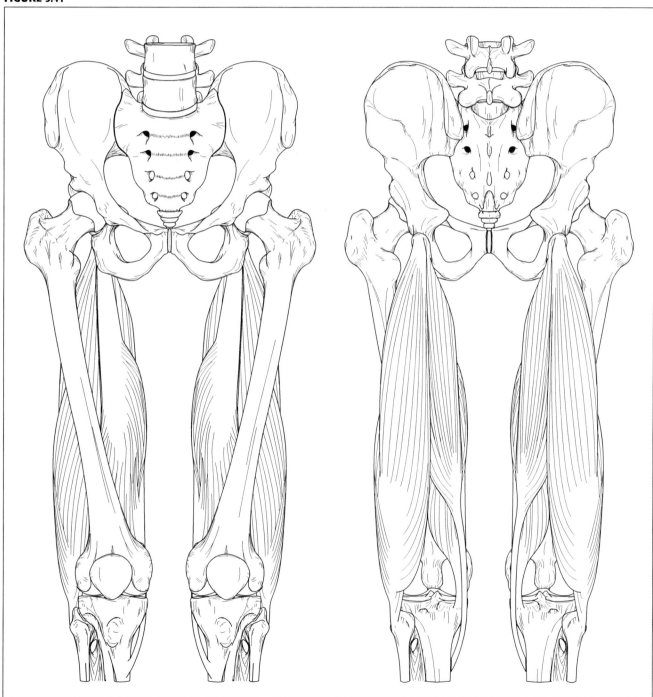

Illustration by Jamey Garbett. © 2003 Mark Nielsen.

Muscles of the Posterior Compartment of the Thigh

Muscle	Origin	Insertion	Function	Innervation
Biceps femoris	Medial aspect of the superior part of the ischial tuberosity and lower portion of the sacrotuberous ligament (long head); lateral edge of lower two-thirds of linea aspera and the lateral supracondylar line (short head)	Lateral and posterior surfaces of the head of the fibula; fibular collateral ligament, and lateral condyle of the tibia	Extension of the hip; flexion of the knee; lateral rotator of the semiflexed knee	Tibial nerve (long head) and common fibular nerve (short head) (L5, S1, and S2)
Semitendinosus	Medial aspect of the superior part of the ischial tuberosity via a shared tendon with the biceps femoris	Medial shaft of the tibia just distal to the medial condyle; attaches slightly distal to the gracilis and posterior to the sartorius	Extension of the hip; flexion of the knee; medial rotator of the semiflexed knee	Tibial nerve (L5, S1, and S2)
Semimem-branosus	Lateral aspect of the superior part of the ischial tuberosity	Tubercle on the posterior surface of medial condyle of the tibia with slips to the medial and lateral borders of the medial condyle	Extension of the hip; flexion of the knee; medial rotator of the semiflexed knee	Tibial nerve (L5, S1, and S2)

Practical applications

Summary of the Thigh Muscles
Anterior View of Complete Hip and Thigh Musculature

FIGURE 5.42

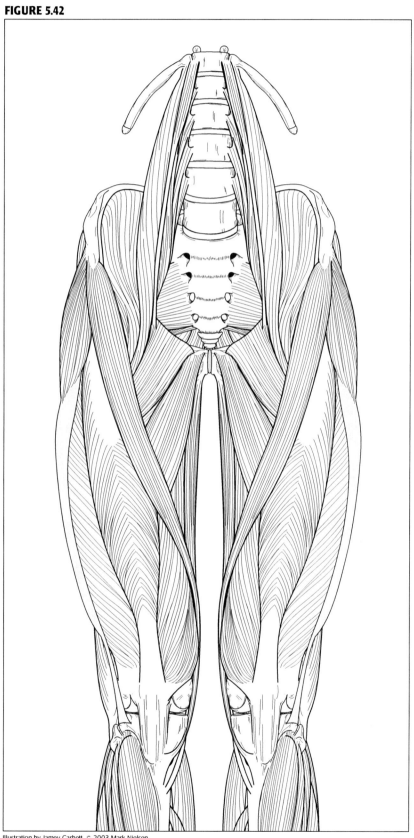

Illustration by Jamey Garbett. © 2003 Mark Nielsen.

Posterior View of Complete Hip and Thigh Musculature

FIGURE 5.43

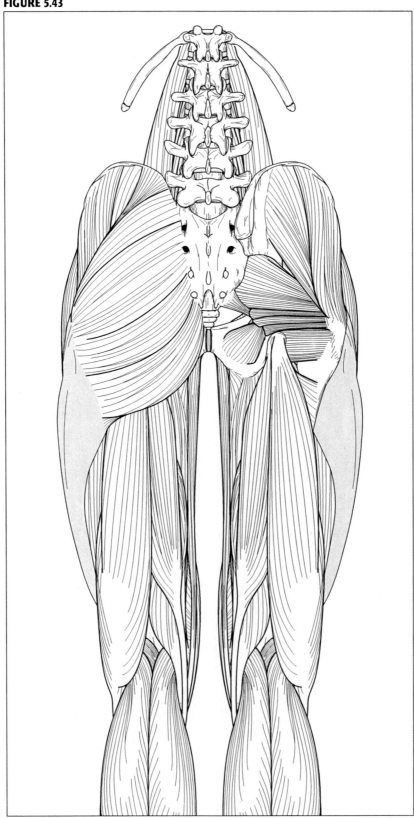

Illustration by Jamey Garbett. © 2003 Mark Nielsen.

Talocrural (Ankle) Joint

Type of Joint

Articulations

Ligaments

Movements of the Ankle Joint

Dorsal flexion

Plantar flexion

Tarsal Joints

Joints

Subtalar joint

Transverse tarsal joint

Talocalcaneonavicular joint

Calcaneocuboid joint

Movements of the Tarsal Joints

Inversion

Eversion

FIGURE 5.44

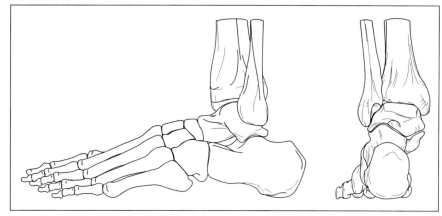

FIGURE 5.45

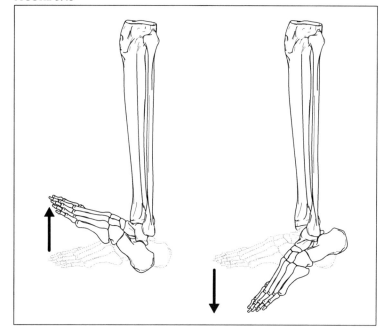

FIGURE 5.46

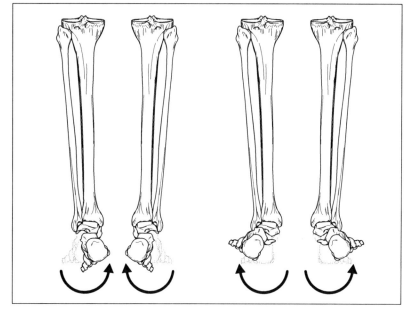

Metatarsophalangeal Joints

Actions

Flexion

Extension

Abduction

Adduction

Interphalangeal Joints

Actions

Flexion

Extension

FIGURE 5.47

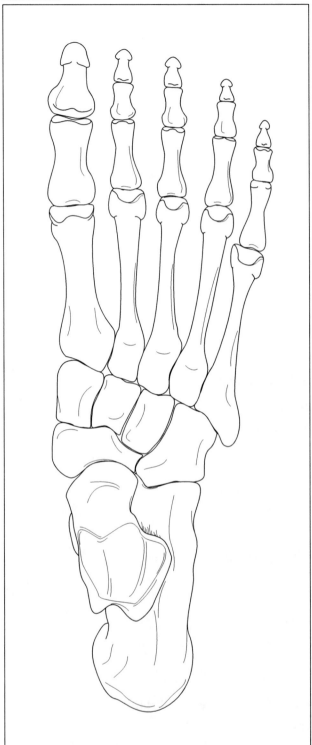

Illustration by Jamey Garbett. © 2003 Mark Nielsen.

Muscles of the Crus

The muscles of the crus cross both the knee and ankle joints as they pass into the foot. Most of these muscles produce their primary actions in the ankle and foot, but some also function at the knee joint.

Anterior Crural Compartment Muscles
Tibialis Anterior Muscle
tibialis = tibia, anterior = in front of

Structure and relationships

Function

Extensor Digitorum Longus Muscle
extensor = one that extends, digitorum = referring to the toes, longus = long

Structure and relationships

Function

FIGURE 5.48

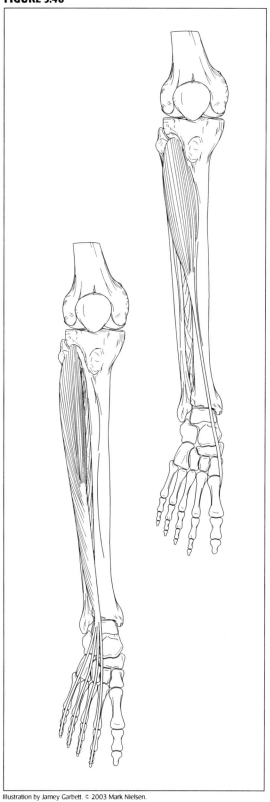

Illustration by Jamey Garbett. © 2003 Mark Nielsen.

Extensor Hallucis Longus Muscle

*extensor = one that extends, **hallux** = big toe, **longus** = long*

Structure and relationships

Function

Fibularis (Peroneus) Tertius Muscle

*peroneus = fibula, **tertius** = third*

Structure and relationships

Function

FIGURE 5.49

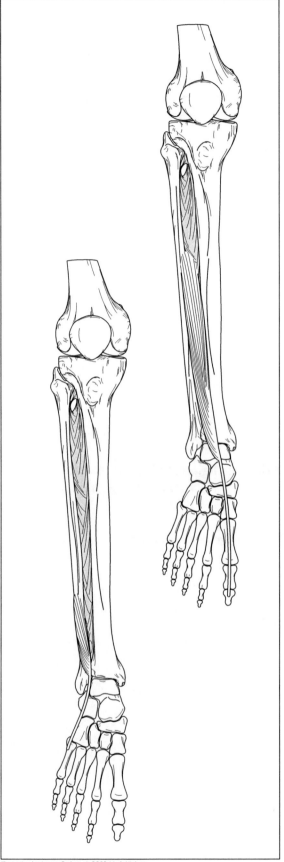

Illustration by Jamey Garbett. © 2003 Mark Nielsen.

Summary of the Anterior Crural Compartment Muscles

Common attachments

Common actions at joints

Practical applications

FIGURE 5.50

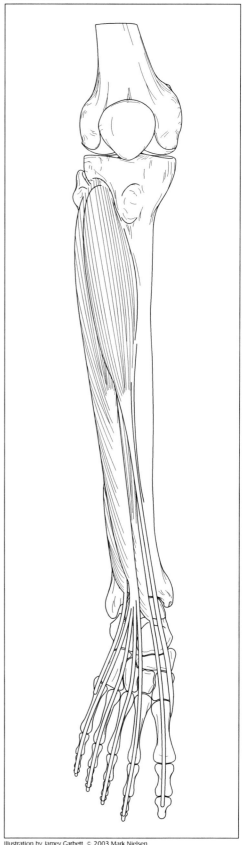

Illustration by Jamey Garbett. © 2003 Mark Nielsen.

Muscles of the Anterior Compartment of the Leg

Muscle	Origin	Insertion	Function	Innervation
Tibialis anterior	Lateral condyle, proximal half of the shaft of the tibia, and adjacent interosseous membrane	Medial and inferior surfaces of the medial cuneiform and the base of the first metatarsal bone	Dorsal flexion of the ankle joint (tibiotalar); inversion of the foot	Deep fibular nerve (L4 and L5)
Extensor digitorum longus	Lateral condyle of the tibia, proximal 3/4ths of the medial surface of the fibula, and adjacent interosseous membrane	Middle phalanx and distal phalanx of the four lesser toes via the extensor expansion	Dorsal flexion of the ankle joint; extension of the four small toes	Deep fibular nerve (L5 and S1)
Extensor hallucis longus	Medial surface of middle portion of the fibula and the adjacent interosseous membrane	Dorsal side of the base of the distal phalanx of the large toe with tendinous expansions to the base of the proximal phalanx on either side	Dorsal flexion of ankle; extension of big toe	Deep fibular nerve (L5)
Fibularis tertius	Medial surface of the distal third of the fibula and adjacent interosseous membrane (fibers blend with the extensor digitorum longus)	Medial side of the dorsal surface of the base of 5th metatarsal and down onto the shaft of this bone	Dorsal flexion of the ankle joint; eversion of the foot	Deep fibular nerve (L5 and S1)

Lateral Crural Compartment Muscles

Fibularis (Peroneus) Longus Muscle

peroneus = fibula, longus = long

Structure and relationships

Function

Fibularis (Peroneus) Brevis Muscle

peroneus = fibula, brevis = short

Structure and relationships

Function

FIGURE 5.51

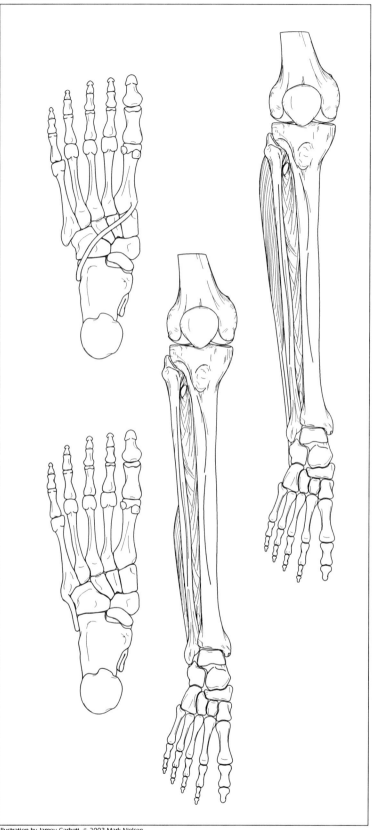

Illustration by Jamey Garbett. © 2003 Mark Nielsen.

Summary of the Lateral Crural Compartment Muscles

Common attachments

Common actions at joints

Practical applications

FIGURE 5.52

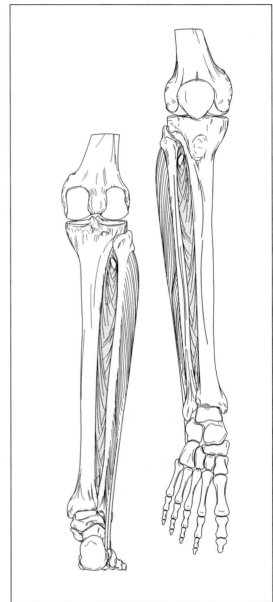

Illustration by Jamey Garbett. © 2003 Mark Nielsen.

Muscles of the Lateral Compartment of the Leg

Muscle	Origin	Insertion	Function	Innervation
Fibularis longus	Lateral surface of head and proximal 2/3rds of the fibula	Lateral side of the base of the first metatarsal bone, lateral side of the base of the medial cuneiform, and the inferior base of the second metatarsal bone	Plantar flexion of the ankle joint; eversion of the foot; helps maintain the longitudinal and transverse arches of the foot	Superficial fibular nerve (L5 and S1)
Fibularis brevis	Lateral surface of the distal 2/3rds of the shaft of the fibula; anterior and posterior crural inter-muscular septa	Tubercle on the lateral base of the 5th metatarsal bone	Weak plantar flexor of the ankle joint; eversion of the foot	Superficial fibular nerve (L5 and S1)

Posterior Crural Compartment Muscles
Deep Group
Tibialis Posterior Muscle
tibialis = tibia, posterior = behind

Structure and relationships

Function

Flexor Digitorum Longus Muscle
flexor = one that flexes, digitorum = referring to the toes, longus = long

Structure and relationships

Function

FIGURE 5.53

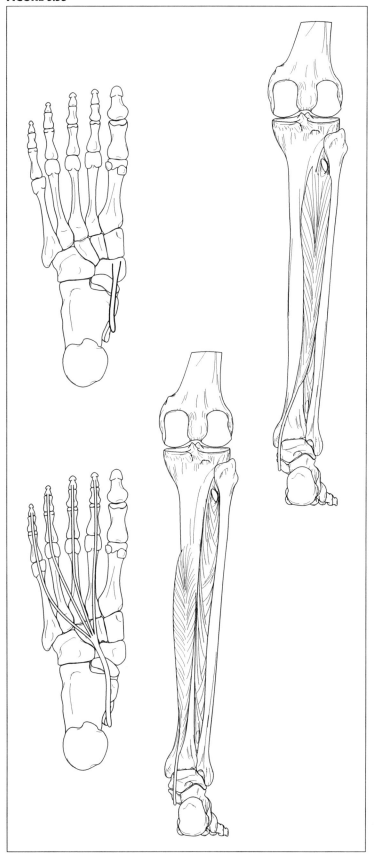

Illustration by Jamey Garbett. © 2003 Mark Nielsen.

Flexor Hallucis Longus Muscle

flexor = one that flexes, *hallux* = big toe,
longus = long

Structure and relationships

Function

FIGURE 5.54

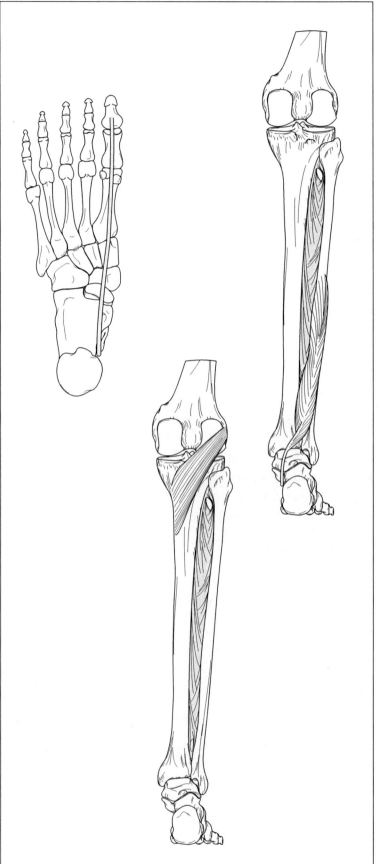

Popliteus Muscle

popliteal = area behind the knee, from *post* =
behind + *plicare* = to fold

Structure and relationships

Function

Illustration by Jamey Garbett. © 2003 Mark Nielsen.

Superficial Group
Triceps Surae Muscle
*tri- = three + **kephale** = head, **sura** = calf of the leg*

Soleus Muscle
solea = sole of the foot

Structure and relationships

Function

Gastrocnemius Muscle
*gastro = belly + **knemia** = leg*

Structure and relationships

Function

FIGURE 5.55

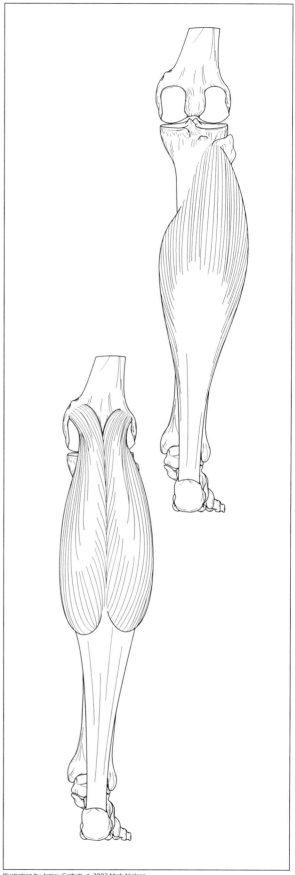

Illustration by Jamey Garbett. © 2003 Mark Nielsen.

Plantaris Muscle

planta = the sole of the foot

Structure and relationships

Function

FIGURE 5.56

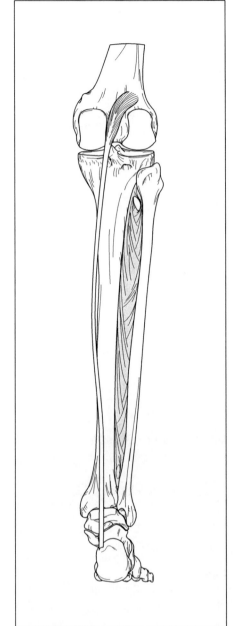

Illustration by Jamey Garbett. © 2003 Mark Nielsen.

Summary of the Posterior Crural Compartment Muscles

Common attachments

Common actions at joints

Practical applications

FIGURE 5.57

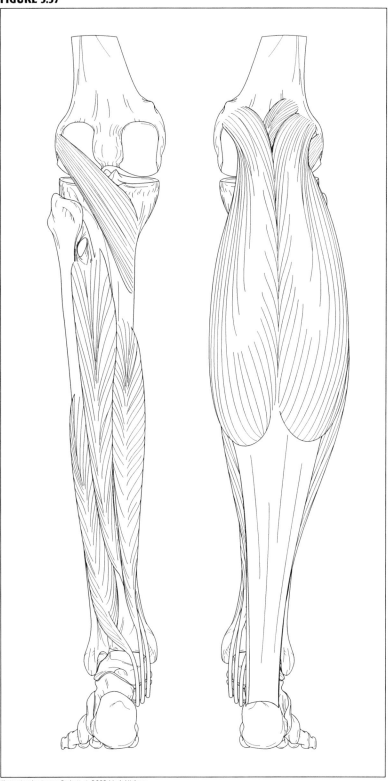

Illustration by Jamey Garbett. © 2003 Mark Nielsen.

Muscles of the Posterior Compartment of the Leg

Muscle	Origin	Insertion	Function	Innervation
Tibialis posterior	Posterior surface of interosseous membrane and the adjacent surfaces of the tibia and the fibula	Tuberosity of the navicular bone, plantar surface of the medial cuneiform bone, tip of the sustentaculum tali, plantar surface of the intermediate cuneiform, and the bases of the second, third, and fourth metatarsal bones	Primary inverter of the foot; assists in plantar flexion of the ankle joint	Tibial nerve (L4 and L5)
Flexor digitorum longus	Posterior surface of the shaft of the tibia from below the soleal line to the level of the proximal end of the medial malleolus	Plantar surface of the base of the distal phalanges of the 4 small toes	Flexion of the little toes; weak plantar flexion of the ankle joint; assist in maintaining the longitudinal arches of the foot	Tibial nerve (L5, S1, and S2)
Flexor hallucis longus	Posterior surface of the distal two-thirds of the fibula, adjacent interosseous membrane, and the fascia of the tibialis posterior	Plantar surface of the base of terminal phalanx of the great toe	Flexion of first toe; weak plantar flexion of the ankle joint; assist in maintaining the longitudinal arches of the foot	Tibial nerve (L5, S1, and S2)
Popliteus	Popliteal groove on the lateral surface of lateral condyle of the femur, lateral side of the lateral meniscus, and the arcuate popliteal ligament of the joint capsule of the knee	Triangular posterior surface of the tibia above the soleal line	Medially rotates tibia on the femur; unlocks the hyperextended knee, retracts the lateral meniscus to prevent it from getting crushed during knee flexion; reduces the load on the posterior cruciate ligament to help prevent anterior dislocation of the femur	Tibial nerve (L4, L5, and S1)
Plantaris	Distal aspect of the lateral supracondylar line and the oblique popliteal ligament	Posterior surface of the calcaneus	Assists with plantar flexion of the ankle joint and flexion of the knee	Tibial nerve (S1 and S2)
Soleus	Posterior surface of the fibular head and proximal end of the fibular shaft, soleal line and medial border of the middle of the tibia	Posterior surface of the calcaneus via the shared tendo calcaneus	Plantar flexion of the ankle joint; plays an important postural role in stabilizing the leg over the ankle in standing	Tibial nerve (S1 and S2)
Gastrocnemius	Proximal surface of lateral condyle of the femur posterior to the adductor tubercle and a ridge on the posterior surface of the femur just proximal to the medial condyle (medial head); lateral surface of the lateral condyle and the lateral supracondylar line (lateral head); both heads send fibers into the posterior surface of the joint capsule of the knee	Posterior surface of the calcaneus via the shared tendo calcaneus	Plantar flexion of the ankle joint; flexion of the knee	Tibial nerve (S1 and S2)

FIGURE 5.58

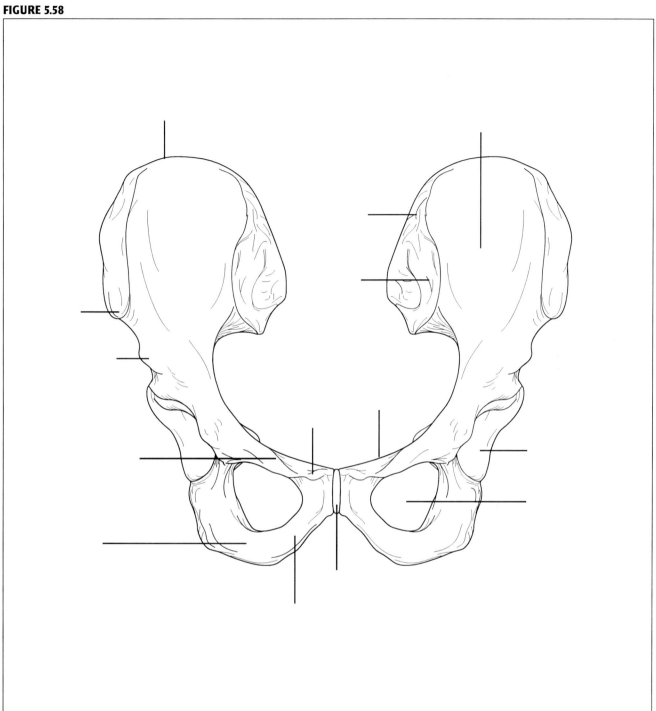

Illustration by Jamey Garbett. © 2003 Mark Nielsen.

FIGURE 5.59

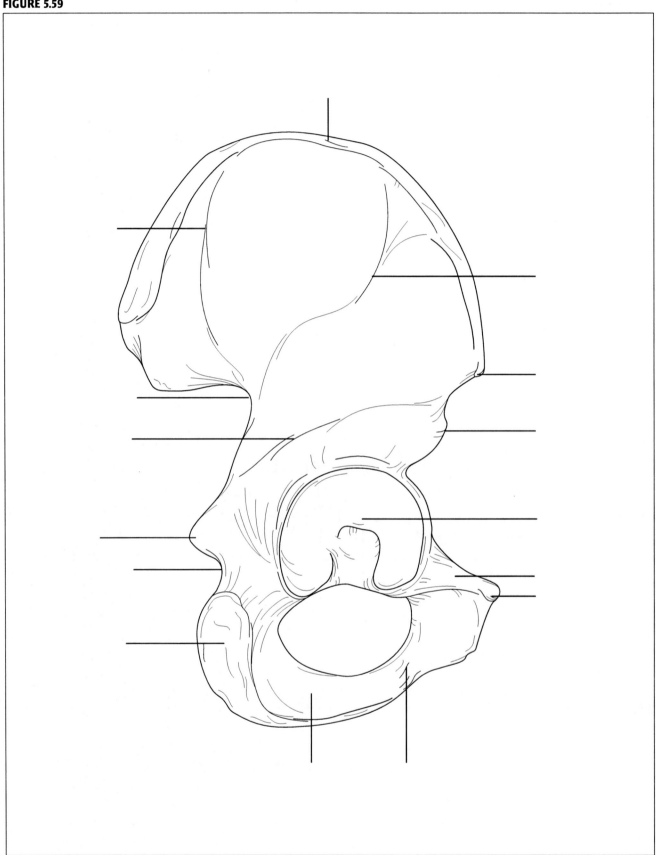

Illustration by Jamey Garbett. © 2003 Mark Nielsen.

FIGURE 5.60

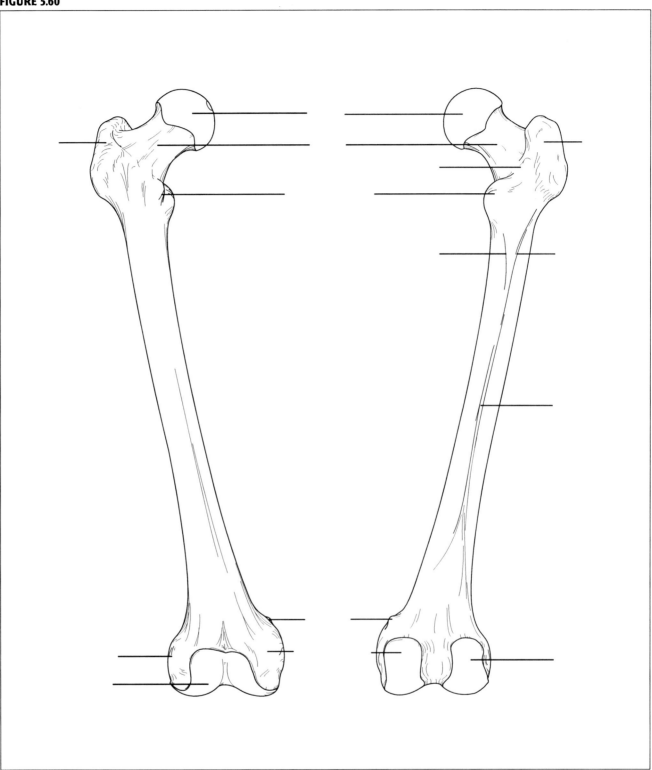

Illustration by Jamey Garbett. © 2003 Mark Nielsen.

FIGURE 5.61

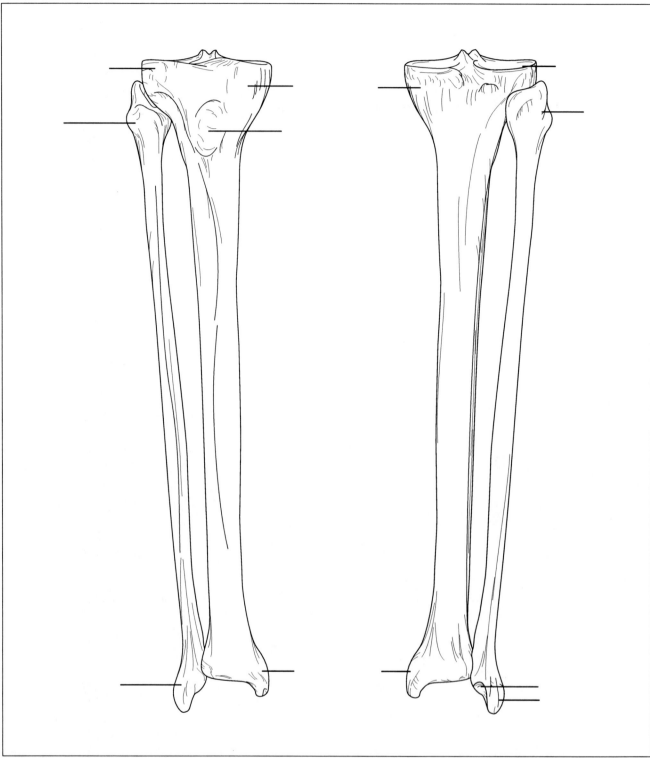

Illustration by Jamey Garbett. © 2003 Mark Nielsen.

FIGURE 5.62

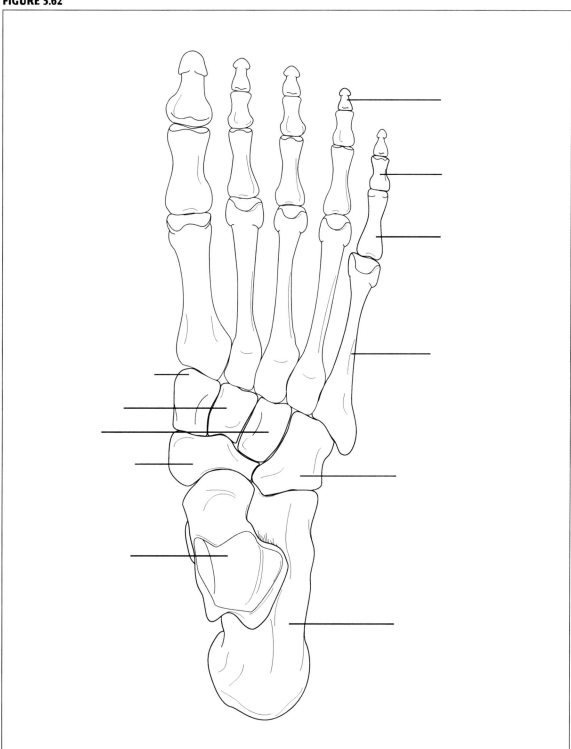

Illustration by Jamey Garbett. © 2003 Mark Nielsen.

Lower Limb

For review, label the muscles shown on the following pages.

FIGURE 5.63

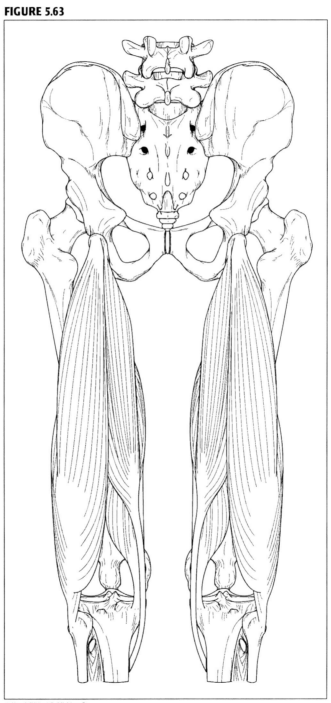

FIGURE 5.64

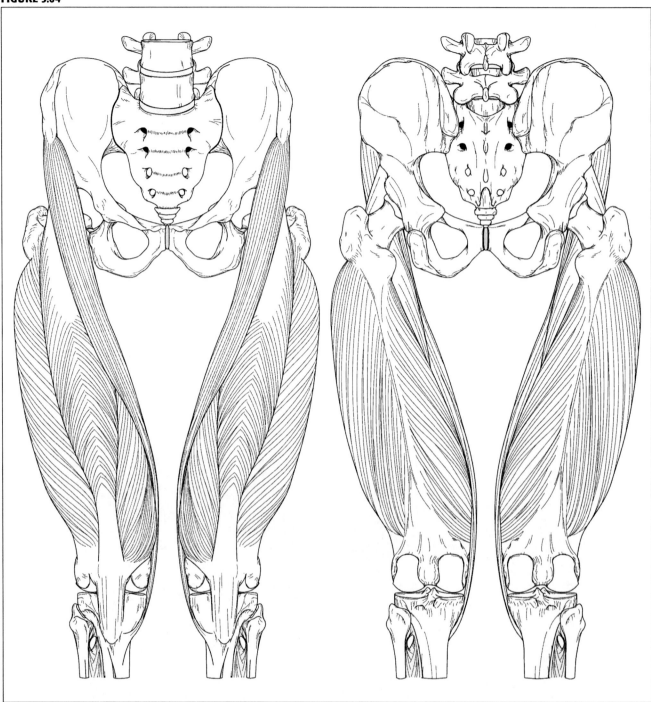

FIGURE 5.65

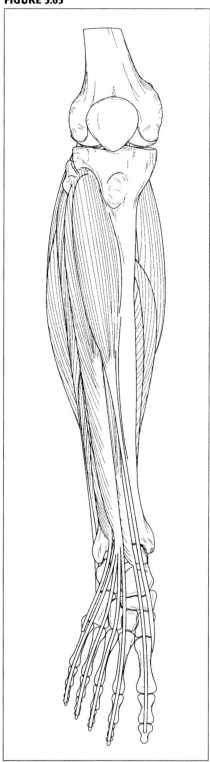

FIGURE 5.66

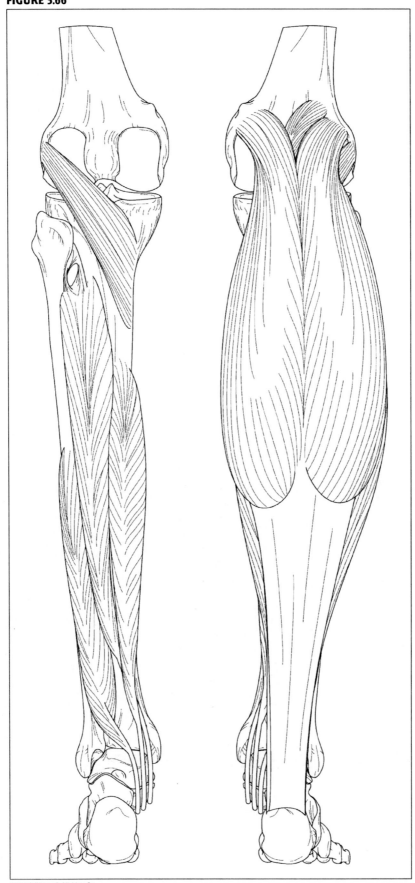

Central Nervous System: The Brain

6

The central nervous system is composed of the brain and spinal cord. It contains interneurons that connect sensory to motor divisions of the peripheral nervous system and glial cells that maintain structural and functional integrity of neuronal structures; it also contains motor neuron cell bodies and the axon terminals of sensory neurons. The brain consists of an interior series of four ventricles filled with cerebrospinal fluid, surrounded by gray and white matter. The protection of these delicate structures is provided by the meninges, cerebrospinal fluid, and bones of the vertebrae and skull.

The adult brain is often described in terms of its five embryonic vesicles: the telencephalon, the diencephalon, the mesencephalon, the metencephalon, and the myelencephalon. This approach will allow the systematic identification of the brain's anatomical features.

OBJECTIVES

☑ Identify the various regions of the brain and their specific landmarks
☑ Know the three layers of the meninges and their spatial arrangement
☑ Identify ventricles or channels associated with each brain region

STRUCTURES YOU ARE RESPONSIBLE FOR IDENTIFYING

Meninges
Dura mater
Arachnoid mater
Subarachnoid space
Pia mater

Telencephalon (Cerebrum)
Sulcus
Gyrus
Hemispheres
Longitudinal fissure
Central sulcus
Precentral gyrus
Postcentral gyrus
Frontal lobe
Parietal lobe
Lateral sulcus (Sylvian fissure)
Temporal lobe
Occipital lobe
Insular lobe
Corpus callosum
Olfactory bulbs
Olfactory nerves/tracts
Basal nuclei

Diencephalon
Thalamus
Hypothalamus
Pituitary (hypophysis)
Mammillary bodies
Pineal body
Optic chiasm
Optic nerve/tracts

Mesencephalon (Midbrain)
Cerebral peduncle
Corpora quadrigemina
Superior colliculus
Inferior colliculus

Metencephalon
Pons
Cerebellum
Arbor vitae
Vermis
Hemispheres

Myelencephalon
Medulla oblongata
Pyramids (of medulla)

Ventricles
Lateral ventricles (inside telencephalon)
Third ventricle (inside diencephalon)
Cerebral aqueduct (inside mesencephalon)
Fourth ventricle (inside metencephalon)
Choroid plexus

Lateral Surfaces of the Brain

The cerebrum is folded into ridges (gyri) and grooves (sulci) allowing a very large surface area of gray matter to fit within the confines of the cranium. The cerebrum has five lobes: the frontal, parietal, temporal, occipital, and insular lobes. The insular lobe can only be seen if the temporal lobe is pulled aside. The diencephalon and mesencephalon cannot be seen from a lateral perspective. The metencephalic structures (cerebellum, pons) can be seen just beneath the cerebral hemispheres. Lastly, the medulla can be seen transitioning into the spinal cord.

Several deep divisions define several regions of the cerebral hemispheres. The two hemispheres are separated by the longitudinal fissure. The central sulcus divides frontal from parietal lobes. The gyri anterior and posterior to this sulcus are called the precentral gyrus and postcentral gyrus, respectively.

FIGURE 6.1

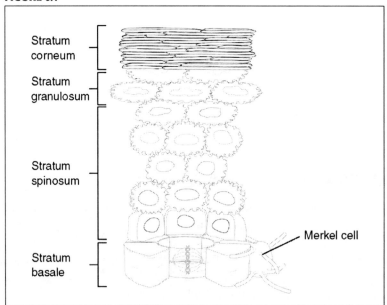

From *Human Anatomy & Physiology* by Stephanie Irwin. Copyright 2006 by Kendall/Hunt Publishing Company. Reprinted by permission.

Sheep Brain, Midsagittal View

From a view of the bisected sheep brain, portions of all major brain regions are visible. The thick, white corpus callosum allows communication between the cerebral hemispheres. The lateral ventricles are found just inferior and lateral to the corpus callosum, one in each hemisphere. The diencephalon surrounds the third ventricle. It is composed of the thalamus (intermediate mass of the sheep brain), pineal body, and hypothalamus. The pituitary gland (hypophysis) hangs from the hypothalamus by a stalk, the infundibulum. The hypophysis is missing in most sheep specimens. A small bump, one half of the mammillary bodies, is visible posterior to the infundibulum. The midbrain or mesencephalon houses a narrow channel known as the cerebral aqueduct. The major landmarks of the mesencephalon are the corpora quadrigemina. This set of four nodules is comprised of a pair of superior colliculi and a pair of inferior colliculi. The pons and cerebellum surround the fourth ventricle, which continues on into the medulla and spinal cord as the central canal.

The cerebellum has a branching structure at its center, known as the arbor vitae.

FIGURE 6.2

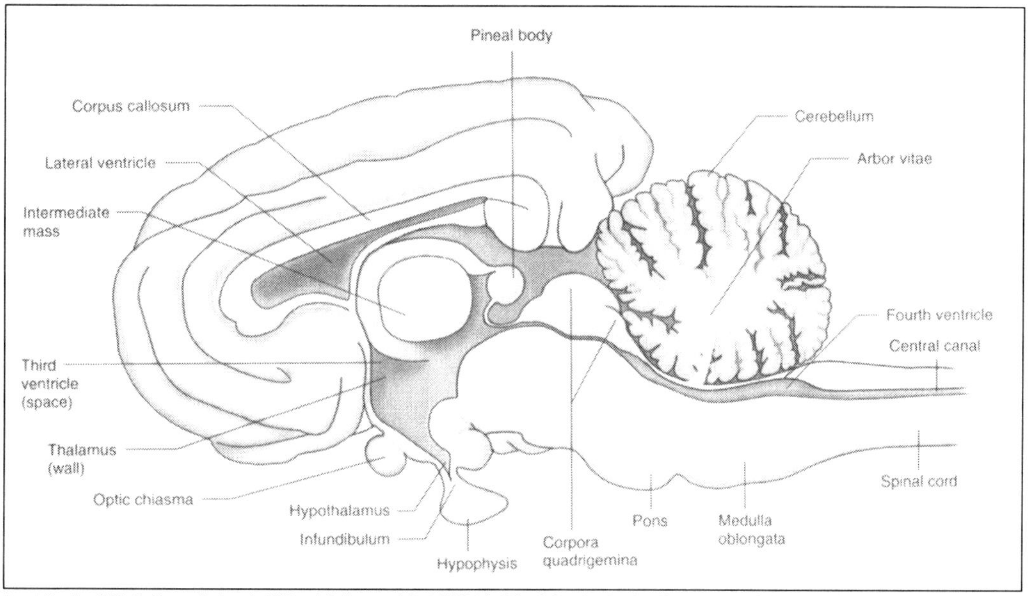

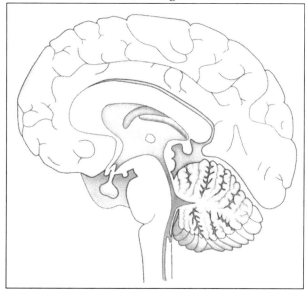

From *Laboratory Guide for Human Anatomy* by William J. Radke, copyright © 2002 John Wiley & Sons, Inc. Reprinted by permission of John Wiley & Sons, Inc.

FIGURE 6.3 Human brain, midsagittal view

From *Laboratory Guide for Human Anatomy* by William J. Radke, copyright © 2002 John Wiley & Sons, Inc. Reprinted by permission of John Wiley & Sons, Inc.

The Heart and Its Circulation

The thoracic cavity is bound anteriorly by the sternum, posteriorly by the vertebral column, laterally and superiorly by the ribs, and inferiorly by the diaphragm. Within the thoracic cavity are three subdivisions: (1) and (2) the right and left lungs, each surrounded by its own pleural cavity; (3) the mediastinum. This latter division contains several structures including (from anterior to posterior): the thymus, the heart within its pericardial sac, the great vessels of the heart, the trachea and bronchi, the esophagus, and the descending aorta.

OBJECTIVES

- ☑ Identify the structures of the heart and its vessels
- ☑ Know the path of blood flow into and out of the heart's four chambers
- ☑ Identify the major arteries leaving the aorta in the thorax
- ☑ Identify the major structures of the respiratory system and understand their functional significance

STRUCTURES YOU ARE RESPONSIBLE FOR IDENTIFYING

Heart
Endocardium
Myocardium
Epicardium
Apex
Right and left atria
Auricles of right and left atria
Right and left ventricles
Interventricular septum
Interatrial septum
Tricuspid (right atrioventricular) valve
Bicuspid or "Mitral" (left atrioventricular) valve
Aortic semilunar valve
Pulmonary semilunar valve
Trabeculae carneae
Papillary muscles
Chordae tendineae
Right and left coronary arteries
Coronary sinus

Major Vessels of the Heart and Thorax
Superior vena cava
Inferior vena cava
Pulmonary trunk
Pulmonary artery
Pulmonary vein
Ascending aorta
Aortic arch
Descending (thoracic) aorta
Brachiocephalic artery (right side only)
Brachiocephalic vein (right and left sides)
Common carotid artery
Internal and external jugular veins
Subclavian artery
Subclavian vein

External Anatomy of the Heart

Each atrium is composed of a principal cavity and a small appendage known as the auricle. The right atrium receives blood from the superior vena cava, the inferior vena cava, and the coronary sinus. The left atrium receives blood from the right and left pulmonary veins.

The right and left ventricles make up most of the heart's mass. The right ventricle pumps blood to the pulmonary trunk, and the left ventricle pumps to the aorta.

The coronary arteries supply blood to thick muscle of the heart, the myocardium. The right and left coronary arteries branch off of the aorta, at its base. Venous blood from the myocardium is collected by a large vein called the coronary sinus, which returns blood to the right atria.

FIGURE 7.1

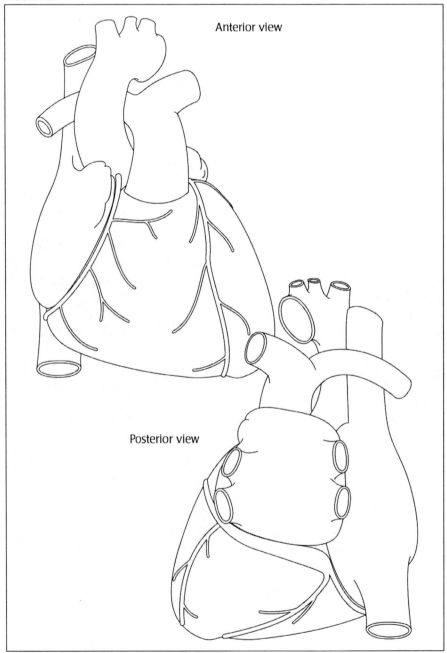

Anterior view

Posterior view

© 2002 Kendall/Hunt Publishing Co.

Internal Anatomy of the Heart

On the internal wall of the atrium, muscular ridges can be seen. These are known as pectinate (pectin = comb) muscles. In the ventricles, these ridges are much larger, forming the trabeculae carneae.

FIGURE 7.2 Internal view of the left atrium and ventricle

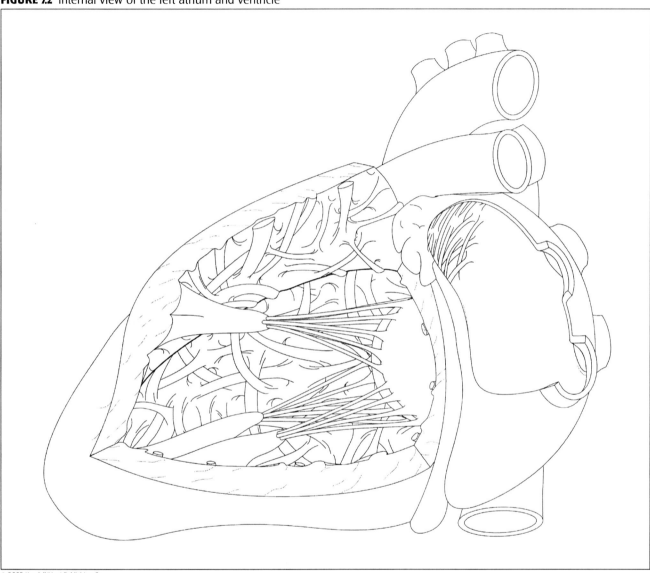

An atrioventricular valve separates the atrium and ventricle on each side of the heart. Valves are composed of three (tricuspid—right side) or two (bicuspid—left side) cusps. By closing as the ventricles contract, the AV valves ensure one-way flow of blood from the ventricle to the arterial system. To ensure that backflow of blood into the atria does not occur, small cords called chordae tendineae anchor the cusps to the walls of the ventricle. Small papillary muscles hold the chordae tendineae in place, contracting with the ventricles.

FIGURE 7.3

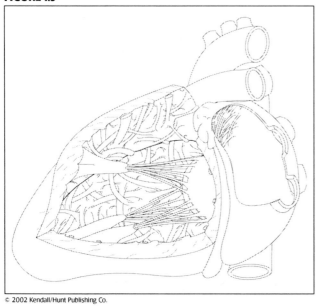

FIGURE 7.4

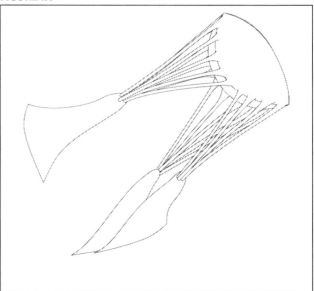

Atrioventricular Valve Function

FIGURE 7.5 Position of AV valves during ventricular filling (left) and ventricular ejection (right).

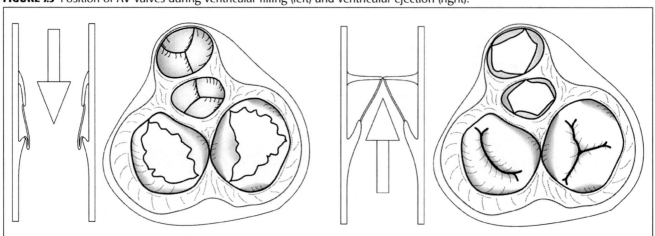

The aorta and pulmonary trunk are separated from their respective ventricles by a semilunar valve. These tricuspid valves prevent backflow of high-pressure blood from the arterial system into the relaxing ventricles. The ostia (openings) of the coronary blood vessels are found just outside of the aortic semilunar valve.

FIGURE 7.6 Aortic semilunar valve with associated coronary arteries

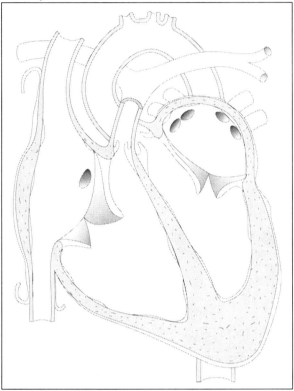

© 2002 Kendall/Hunt Publishing Co.

FIGURE 7.7 Aorta has been opened to display three cusps of its valve

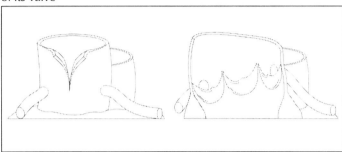

© 2002 Kendall/Hunt Publishing Co.

Semilunar Valve Function

FIGURE 7.8 Position of semilunar valves during ventricular filling (left) and ventricular ejection (right)

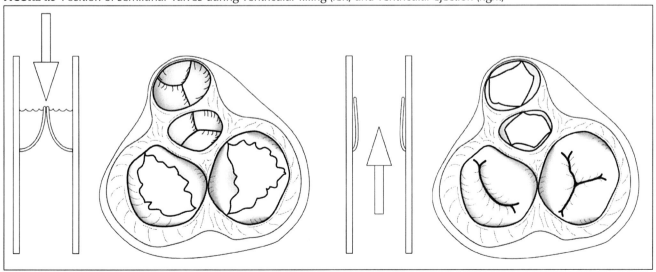

© 2002 Kendall/Hunt Publishing Co.

Functional Anatomy of the Heart

The flow of blood in the body occurs through two major circuits, the systemic and the pulmonary. Blood enters the right atrium from all tissues of the body. This oxygen-poor blood flows into the right ventricle and is pumped to the lungs via the pulmonary trunk and arteries. After picking up oxygen, blood returns to the left atrium where it then flows into the left ventricle. The left ventricle sends blood to the entire body via the aorta.

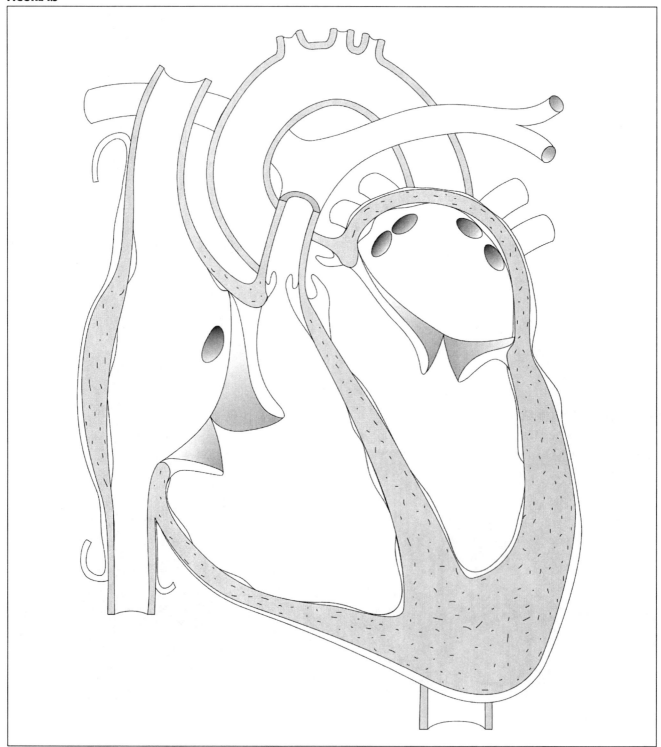

Major Arteries of the Body

The aorta distributes oxygenated blood to the entire body. The major branches of the aorta are shown below.

FIGURE 7.10

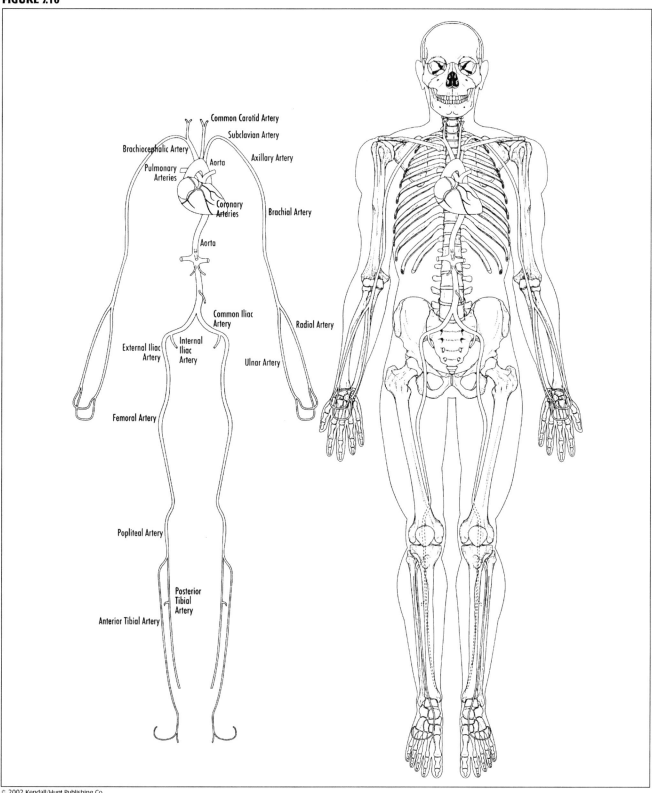

The Heart and Its Circulation

For review, label the illustrations below.

Apex	Left auricle	Inferior vena cava	Aorta
Right atrium	Right ventricle	Pulmonary trunk	Brachiocephalic artery
Left atrium	Left ventricle	Pulmonary artery	Common carotid artery
Right auricle	Superior vena cava	Pulmonary vein	Subclavian artery

FIGURE 7.11 Anterior view of external heart

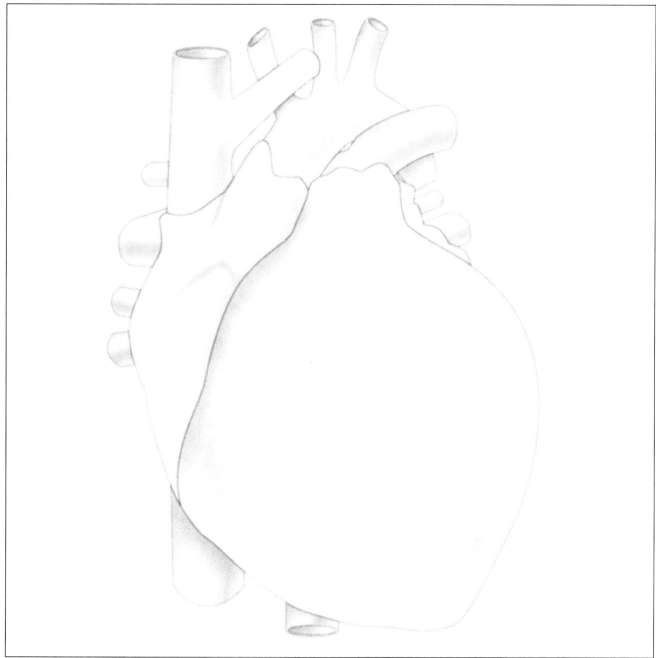

FIGURE 7.12 Posterior view of external heart

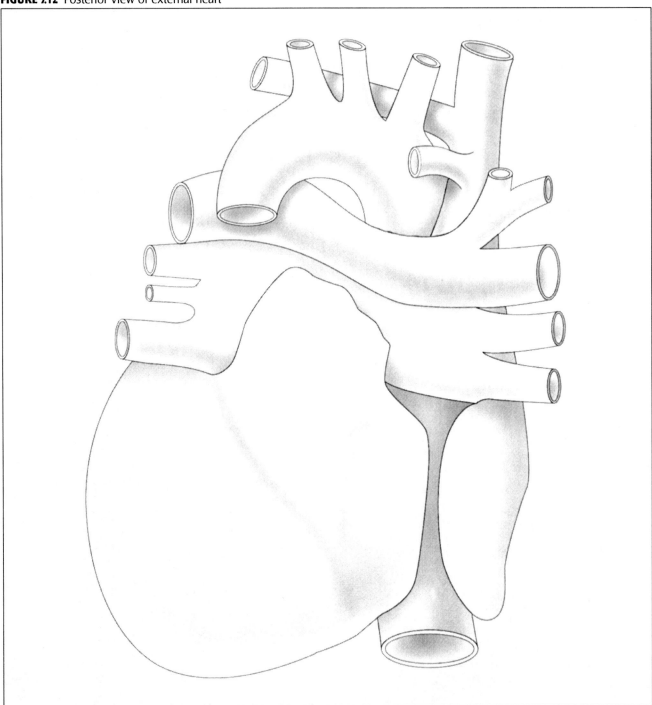

The Respiratory System

The respiratory system extends from the external nares (nostrils) to the lungs. Its primary responsibilities are to transport air into and out of the alveoli where gas exchange occurs. Secondarily, the respiratory system is involved in phonation and acid-base balance of the blood.

OBJECTIVES

- ☑ Understand the structures involved in the passage of air from the external nares to the lungs
- ☑ Identify and distinguish between primary, secondary, and tertiary bronchi
- ☑ Know the structure of the larynx and how sound is produced
- ☑ Understand the arrangement of pleural membranes and the pleural cavities

STRUCTURES YOU ARE RESPONSIBLE FOR IDENTIFYING

External nares
Nasal cavity
   Nasal conchae
Oral cavity
Paranasal sinuses
Hard palate
Soft palate
Pharynx
   Nasopharynx
   Oropharynx
   Laryngopharynx
Larynx
   Thyroid cartilage
   Cricoid cartilage
   Epiglottis

Arytenoid cartilages
Vocal folds
Glottis
Right and left lungs
   Superior lobe
   Inferior lobe
   Middle lobe (right lung only)
Trachea
Primary bronchus
Secondary bronchus
Tertiary bronchus
Parietal pleura
Visceral pleura
Pleural cavity
Diaphragm

Overview of the Respiratory Tract

The portion of the respiratory system that functions to transmit air to and from the respiratory zone is called the conduction zone. This passageway begins with the nasal cavity, passes through the three divisions of the pharynx (nasopharynx, oropharynx, and laryngopharynx) and into the larynx before entering the respiratory tree. The respiratory tree consists of the trachea and the primary, secondary, and tertiary bronchi. The two primary bronchi enter the lungs. The secondary (lobar) bronchi enter each lobe of the lungs, and the tertiary (segmental) bronchi enter the bronchopulmonary segments that form functional units of the lung supplied by a distinct artery and vein.

FIGURE 8.1

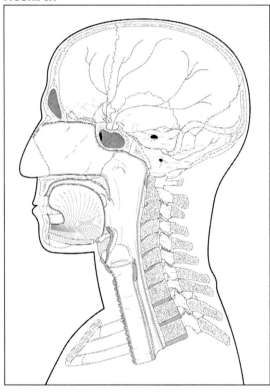

FIGURE 8.2

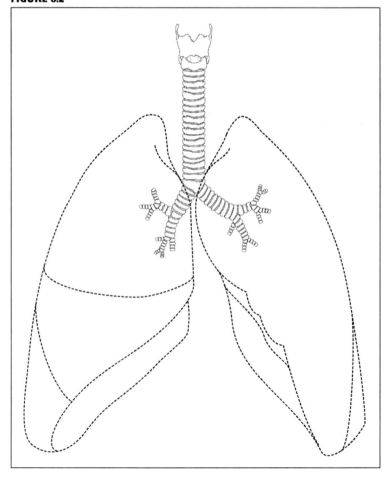

Illustrations by Jamey Garbett. © 2003 Mark Nielsen.

Structure of the Larynx

The larynx stands between the trachea and the pharynx. To prevent food and water from entering the lower respiratory passages, the epiglottis serves as a valve to cover the superior opening of the larynx. Two other cartilages form the skeleton of the larynx: the large thyroid cartilage (Adam's apple) and the smaller cricoid cartilage beneath. Within the larynx, folds of the mucosal lining are brought into contact. These vocal folds (cords) are positioned by a pair of arytenoid cartilages, ligaments stretched across the thyroid cartilage, and several small muscles. The vocal folds are vibrated by expelled air to produce sounds. Additionally, vocal folds are sometimes held together to prevent the passage of air for the purpose of increasing intra-abdominal pressure, such as just prior to coughing.

The Cartilages of the Larynx

FIGURE 8.3 Anterior (left) and posterior views (middle) of the larynx; midsagittal view of the left half of the larynx (right)

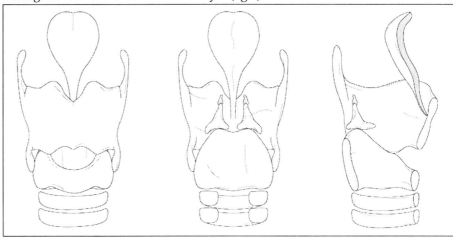

Illustration by Jamey Garbett. © 2003 Mark Nielsen.

Function of the Arytenoid Cartilages and Associated Muscles

FIGURE 8.4

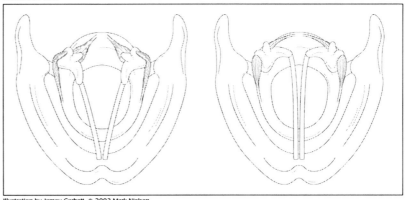

Illustration by Jamey Garbett. © 2003 Mark Nielsen.

Cartilages, Muscles, and Ligaments lined by Mucosa Forming the Vocal Folds

FIGURE 8.5

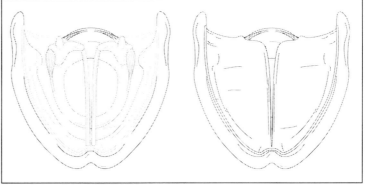

Illustration by Jamey Garbett. © 2003 Mark Nielsen.

The Divisions of the Thoracic Cavity

Within the thorax, there are several divisions. On the right and left, the lungs sit within the pleural cavities. These cavities are surrounded by the parietal pleura, which lines the thoracic wall, and the visceral pleura, which covers the lungs. In the center of the thorax, the mediastinum contains a number of structures including the heart in its pericardial cavity, the trachea and primary bronchi, the esophagus, and the aorta and its major branches.

FIGURE 8.6

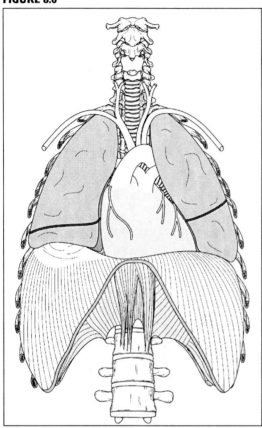

FIGURE 8.7

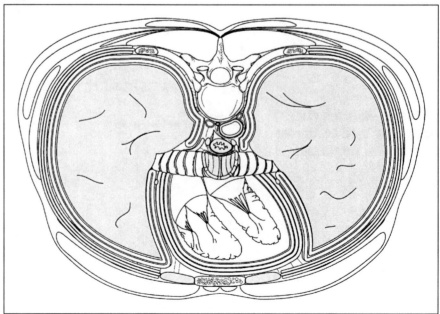

The Respiratory System

For review, identify the membranes of the lungs and heart, and identify as many other structures as you can in the figures below.

FIGURE 8.8

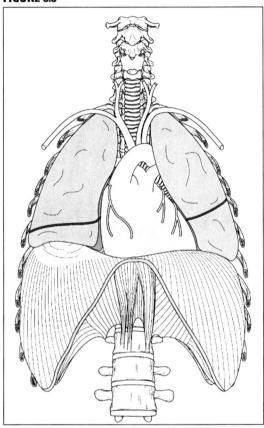

FIGURE 8.9

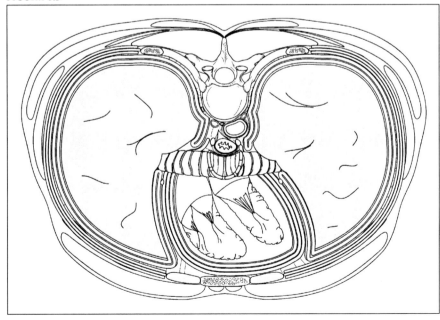

Digestive Organs

9

The abdominopelvic cavity extends from the diaphragm to the floor of the pelvis and contains two divisions. The larger and upper division is the abdominal cavity. This cavity is bound by the abdominal wall and the lumbar vertebrae. The smaller and lower division is the pelvic cavity, which houses the organs of the urinary and reproductive systems and is bound by the bones of the pelvis and sacrum.

OBJECTIVES

- ☑ Identify the digestive and accessory digestive organs and their major anatomical features
- ☑ Identify the blood vessels that serve the abdominal organs

STRUCTURES YOU ARE RESPONSIBLE FOR IDENTIFYING

Oral cavity
Hard palate
Soft palate
Salivary glands
Oropharynx
Laryngopharynx
Esophagus
Parietal peritoneum
Visceral peritoneum
Mesentery
Stomach
  Cardiac region
  Fundus
  Body

Pylorus
Greater curvature
Lesser curvature
Rugae
Greater omentum
Small intestine
  Duodenum
  Jejunum
  Ileum
  Ileocecal junction/valve
Large intestine
  Cecum
  Appendix
  Ascending colon

Transverse colon
Descending colon
Sigmoid colon
Rectum
Anus
Liver
Gallbladder
Pancreas
Spleen (note: not a digestive organ)
Blood vessels
  Celiac trunk
  Superior mesenteric artery
  Inferior mesenteric artery
  Hepatic portal vein

Structure of the Stomach

The esophagus opens to the stomach at the cardia. Superior to the cardia is the rounded roof of the stomach, the fundus. The body of the stomach tapers towards the small intestine as the pylorus. The pyloric sphincter is a strong muscle that gates entrance to the duodenum of the small intestine. Inside the stomach, the mucosa forms large folds called rugae that flatten as the stomach expands with food.

FIGURE 9.1 Stomach with interior revealed

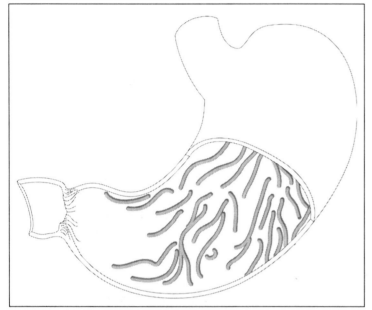

FIGURE 9.2 Digestive organs in situ

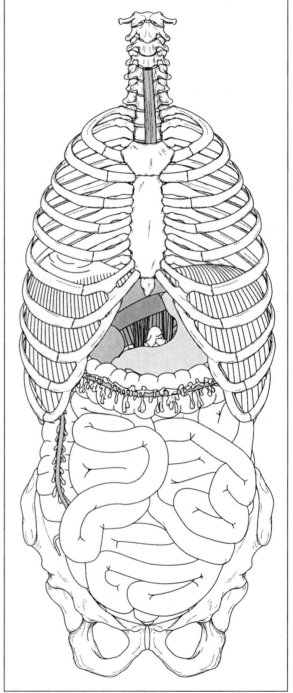

Illustrations by Jamey Garbett. © 2003 Mark Nielsen.

The Small Intestine and Accessory Organs

The pyloric sphincter allows chyme to enter the duodenum. This region is a small (10") C-shaped segment that curls around the pancreas and is the first segment of the small intestine. Here, bile from the liver and gall bladder enters via the bile duct. The secretions of the pancreas also enter at the duodenum. The relatively short duodenum turns into the jejunum, which occupies the upper left quadrant of the abdomen. The lower right quadrant is occupied by the third segment, the ileum.

FIGURE 9.3 Liver, gall bladder, and pancreas and their connection to the duodenum

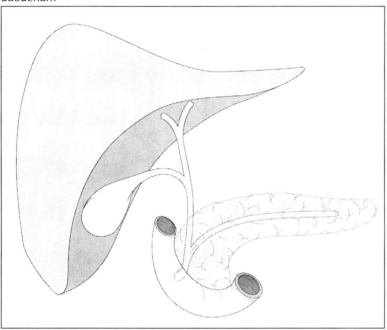

FIGURE 9.4 Digestive organs in situ

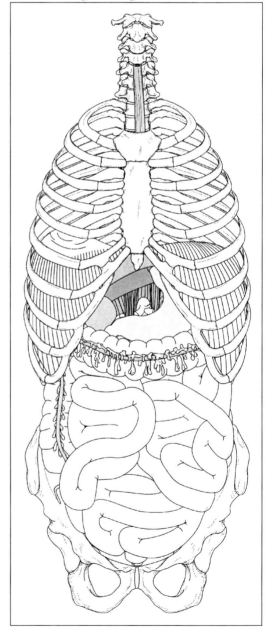

Illustrations by Jamey Garbett. © 2003 Mark Nielsen.

Structure of the Large Intestine

The ileum empties into the cecum of the large intestine at the ileocecal junction. The cecum is a small sac to which attaches the appendix. Material from the cecum is passed up, over, and down the ascending, transverse, and descending colon. From there it is passed into the sigmoid colon, which turns into the pelvic cavity and becomes the rectum.

FIGURE 9.5 The large intestine

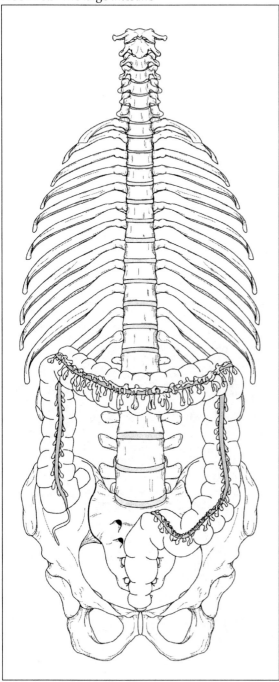

Illustration by Jamey Garbett. © 2003 Mark Nielsen.

FIGURE 9.6 Digestive organs in situ

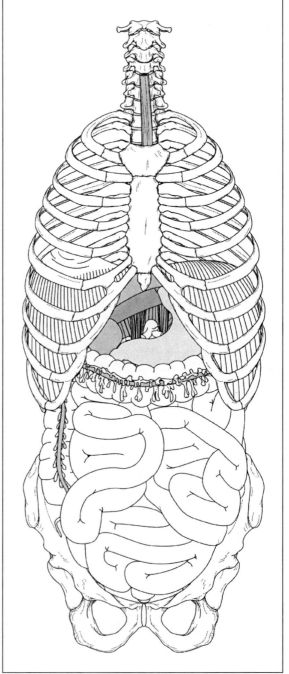

Illustration by Jamey Garbett. © 2003 Mark Nielsen.

The Peritoneal Cavity

The peritoneal cavity surrounds most of the digestive viscera. It is formed by the visceral and parietal peritoneum, connected posteriorly by the mesentery. The mesentery transmits blood vessels and nerves to the viscera and maintains their relative position within the cavity. Anteriorly, the greater omentum covers much of the large and small intestines. This sheet of peritoneal membrane is a storage site for fat and reduces friction as the visceral and body wall come in contact.

FIGURE 9.7 Parietal peritoneum, visceral peritoneum, and mesentery border the peritoneal cavity

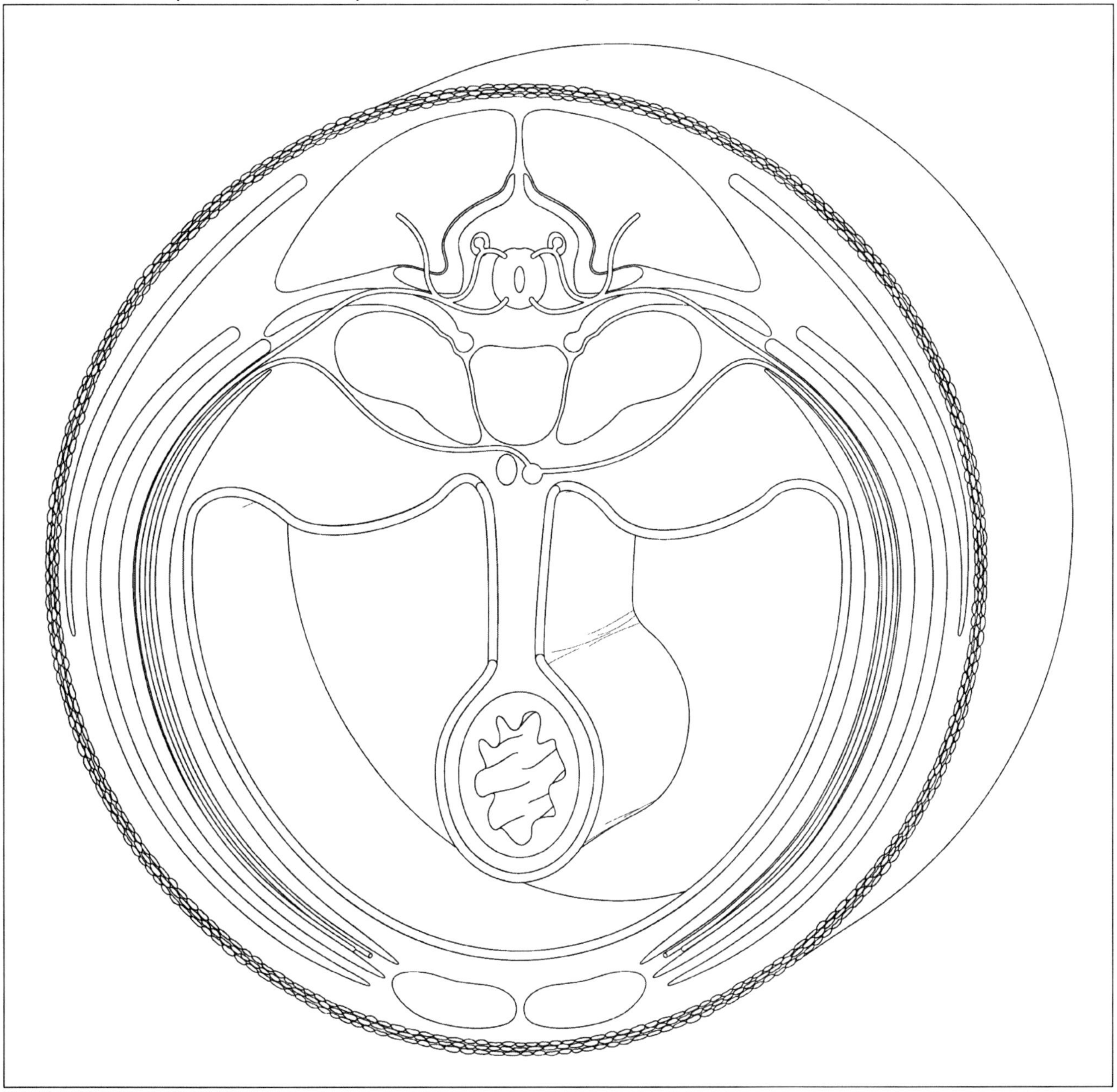

Illustration by Jamey Garbett. © 2003 Mark Nielsen.

Cat Prosection: Abdominal Organs

Because the cat bears such similarity to the human in its abdominopelvic anatomy, several cats have been placed around the lab. These previously dissected (prosected) specimens are, in some respects, smaller versions of human anatomy. The illustrations that follow diagram the cat in order to facilitate the identification of the major organs of the abdomen and pelvis.

FIGURE 9.8 Abdominal organs of the cat. Small intestine reflected to the left to visualize remaining organs. Note: The cat has a very small or no appendix. Also note that the labeled ileocolic sphincter is equivalent to the ileocecal sphincter.

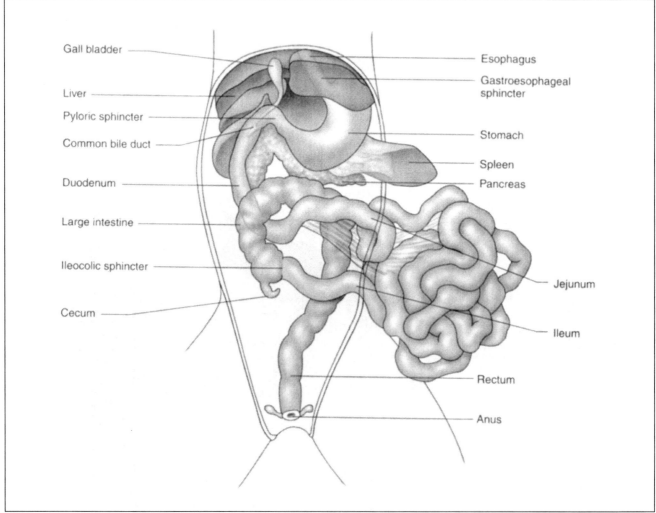

From *Laboratory Guide for Human Anatomy* by William J. Radke, copyright © 2002 John Wiley & Sons, Inc. Reprinted by permission of John Wiley & Sons, Inc.

Digestive Organs

For review, label the illustrations below.

FIGURE 9.9

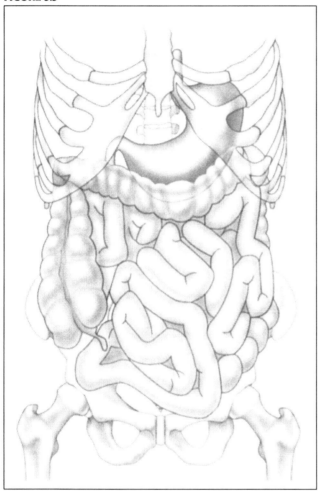

FIGURE 9.10

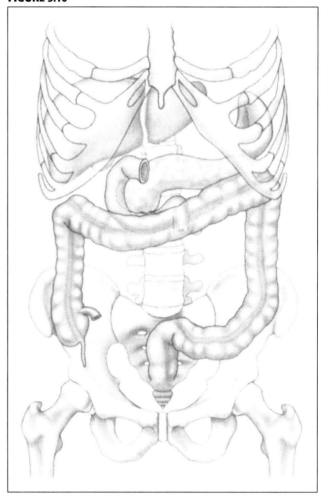

Urinary and Reproductive Organs

10

The urinary and reproductive organs are found outside of the peritoneal cavity. The kidneys are found posterior to the peritoneal cavity, within the abdomen. The reproductive organs are housed in the pelvic cavity, along with the urinary bladder and rectum.

OBJECTIVES

☑ Identify the organs of the urinary system and their blood supply
☑ Identify the reproductive organs and their major anatomical features, including the external genitalia

STRUCTURES YOU ARE RESPONSIBLE FOR IDENTIFYING

Kidney
   Renal capsule
   Medullary region
   Renal pyramids
   Renal columns
   Renal pelvis
Renal arteries
Renal veins
Ureters
Urinary bladder
Urethra
   Prostatic urethra
   Membranous urethra
   Spongy urethra
Penis
   Glans
   Corpus spongiosum
   Corpora cavernosa

External urethral orifice
Urogenital sinus (cat only)
Scrotum
Epididymis
Ductus (vas) deferens
Prostate gland
Seminal vesicles
Ejaculatory ducts
Testis
Labia majora
Labia minora
Clitoris
Vagina
Uterus
Cervix
Uterine (fallopian) tubes
Ovary

Overview of the Urinary System

The kidneys filter blood from the renal arteries and produce urine as waste. Urine is transmitted by the ureters to the urinary bladder. Urine passes out of the body from the bladder through the urethra.

FIGURE 10.1

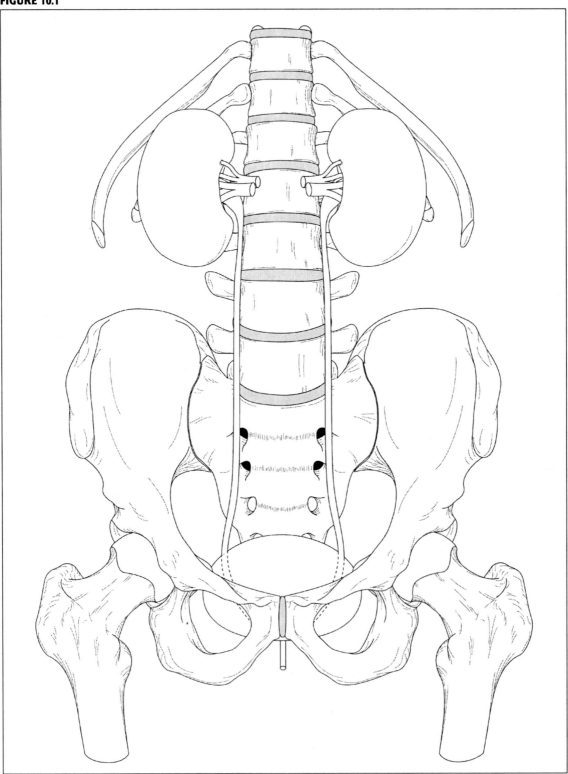

Illustration by Jamey Garbett. © 2003 Mark Nielsen.

The Kidneys

The kidneys are covered by a thin but tough renal capsule. Just beneath, the outer portion of the kidney is called the cortex, while the inner region is the medulla. The medulla is made of discrete segments, the renal pyramids, which contain collecting ducts that drain urine to the minor calyces. Minor calyces converge to form major calyces, which in turn, converge to form the renal pelvis. This is the portion of the kidney that narrows to become the ureter.

FIGURE 10.2

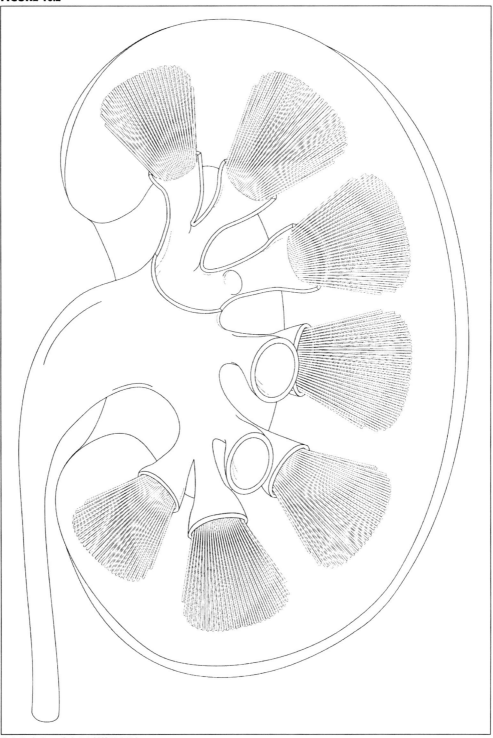

Illustration by Jamey Garbett. © 2003 Mark Nielsen.

The Pelvic Cavity

The line between the abdominal and pelvic cavities is an imaginary one, but is generally considered to lie at the pelvic brim (inlet). This roughly corresponds to the inferior margin of the peritoneal cavity. Thus, all pelvic viscera are outside of and inferior to the peritoneal cavity.

In the male, the major structures occupying the pelvic cavity are the urinary bladder, the prostate (surrounding the urethra), and the rectum. In the female, the uterus, uterine tubes, and ovaries are present, while the prostate is absent.

FIGURE 10.3 Male pelvic organs

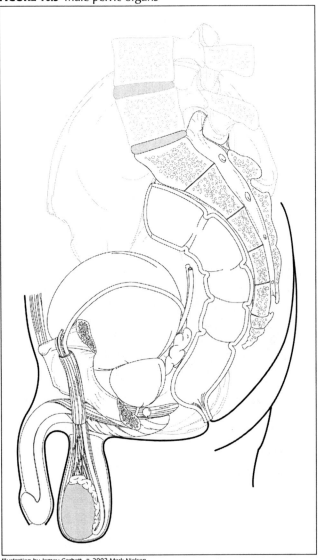

FIGURE 10.4 Female pelvic organs

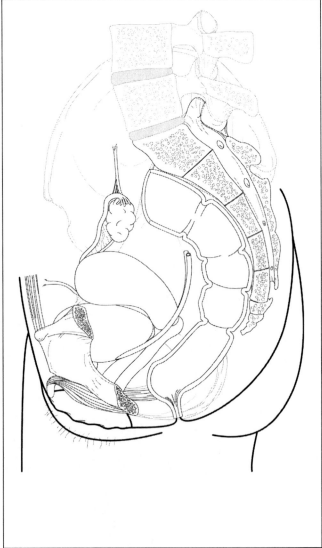

The Male Reproductive System

The testes and vas deferens lie within the scrotum, outside of the pelvic cavity. The vas deferens, however, travel through the anterior abdominal wall and into the pelvic cavity where they come to lie on the posterior wall of the bladder. Here, they join with the paired seminal vesicles to form paired ejaculatory ducts. These ducts empty into the prostate, allowing semen to pass into the urethra. The urethra has three regions throughout its course. As it passes through the prostate, it is called the prostatic urethra. Then, as it passes through the muscular floor of the pelvis, it is known as the membranous urethra. Here, the small paired bulbourethral glands empty their contents into the urethra. Finally, the urethra passes out of the pelvis and through the penis, where it is named the spongy (or penile) urethra.

FIGURE 10.5

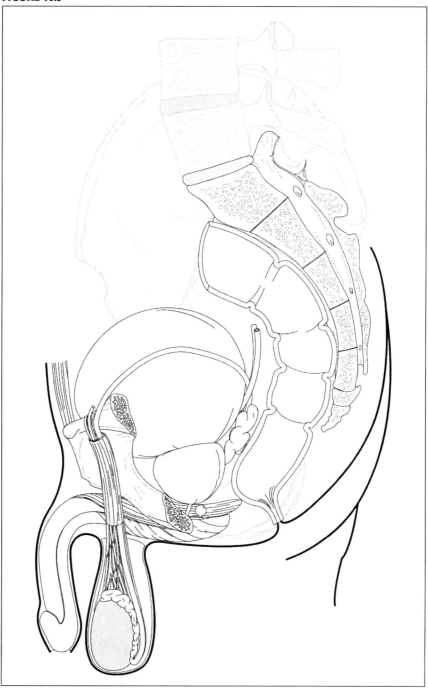

Illustration by Jamey Garbett. © 2003 Mark Nielsen.

FIGURE 10.6 Ducts of male leading from the vas deferens to the urethra

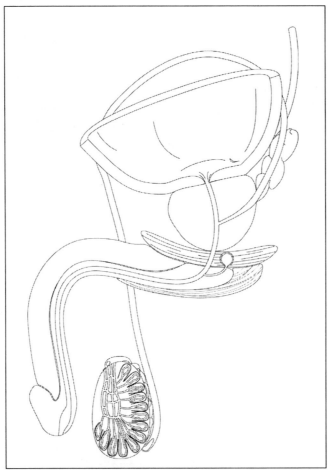

Illustration by Jamey Garbett. © 2003 Mark Nielsen.

FIGURE 10.7 Posterior view of urinary bladder, showing the three accessory glands of the male: prostate, seminal vesicles, and bulbourethral glands

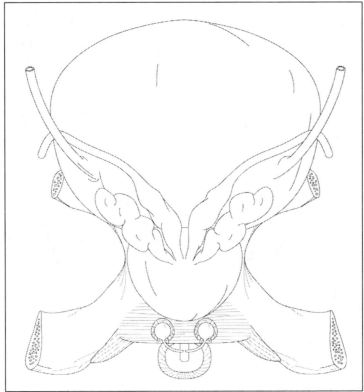

Illustration by Jamey Garbett. © 2003 Mark Nielsen.

The Male External Genitalia

The penis is composed largely of erectile tissue, divided into three chambers. The paired upper chambers are the corpora cavernosa. They remain in close contact throughout the length of the penis, but at its base, they diverge to attach to the inferior rami of the pubic bones. The third chamber, the corpus spongiosum, surrounds the urethra as it emerges from the pelvis. It also forms the glans of the penis. The scrotum lies posterior to the penis and houses the testes. The testes are kept slightly cooler than body temperature by the contraction or relaxation of the thin muscular wall of the scrotum, the cremaster muscle.

FIGURE 10.8 Cross section through penis

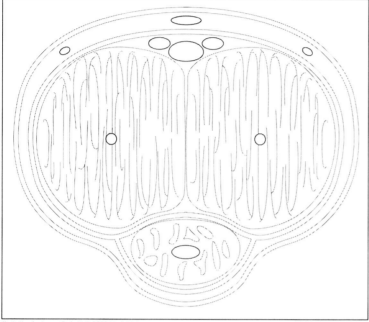

Illustration by Jamey Garbett. © 2003 Mark Nielsen.

FIGURE 10.9 Anterior (left) and posterior (right) views of the corpora cavernosa, showing attachment to the pubic bones

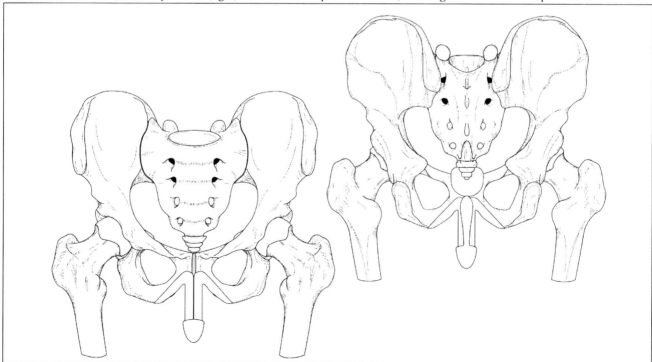

Illustration by Jamey Garbett. © 2003 Mark Nielsen.

The Female Reproductive System

The ovaries, uterine tubes, and uterus lay between the urinary bladder and the rectum within the pelvic cavity. The uterus is connected to the exterior by the vagina, a muscular tube that serves as both the birth canal and the organ of sexual intercourse.

FIGURE 10.10

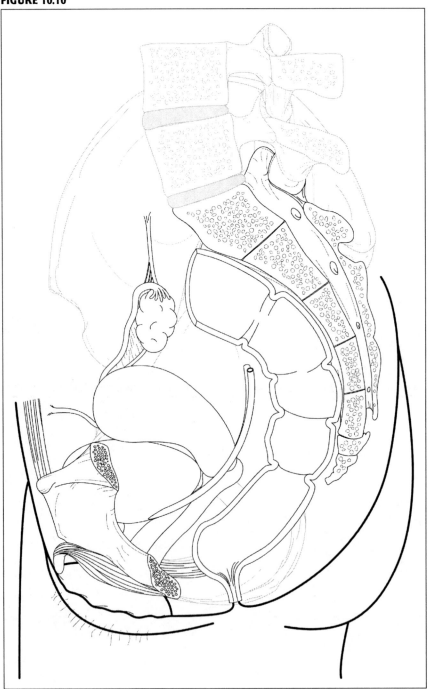

Illustration by Jamey Garbett. © 2003 Mark Nielsen.

The Female External Genitalia

The external genitalia of the female comprise the vulva. This area is bound by the skin-covered labia majora. Just internal are the mucosa-covered labia minora, which merge anteriorly to form the hood of the clitoris. The clitoris is a small erectile organ derived from a common embryological structure with the penis. Posterior to the clitoris is the opening of the urethra, followed by that of the vagina.

FIGURE 10.11

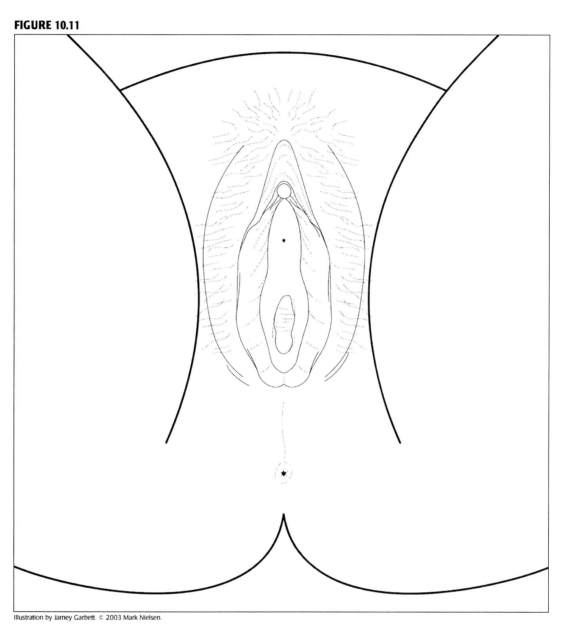

Cat Prosection: Male Urogenital Organs

The male cat differs only slightly from the human in its urogenital system anatomy. In the human, the prostate is directly inferior to the urinary bladder. In the cat, however, the prostate is a few centimeters below the bladder. Note the position of the vas deferens relative to the ureters. This is typical of all mammals with external testes.

FIGURE 10.12 Male cat urogenital anatomy

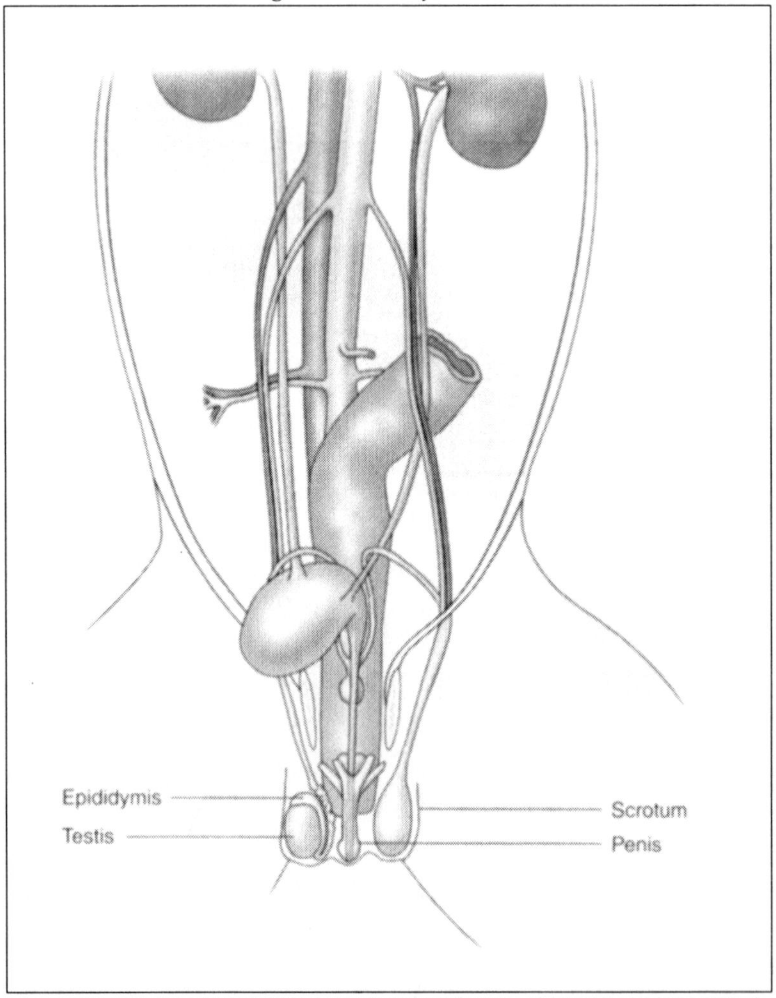

From *Laboratory Guide for Human Anatomy* by William J. Radke, copyright © 2002 John Wiley & Sons, Inc. Reprinted by permission of John Wiley & Sons, Inc.

Cat Prosection: Female Urogenital Organs

The female cat differs only slightly from the human in its urogenital system anatomy. In the human, the urethra and the vagina exit the body separately. In the cat, however, these two tubes combine just before leaving the body, forming a short chamber known as the urogenital sinus.

FIGURE 10.13 Female cat urogenital anatomy

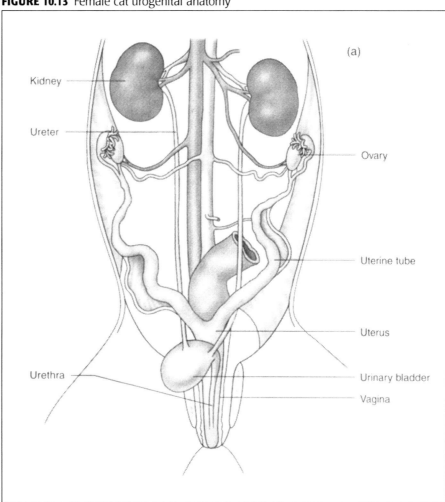

Urinary and Reproductive Organs

For review, label the illustrations below.

FIGURE 10.14 Male urogenital system, midsagittal section

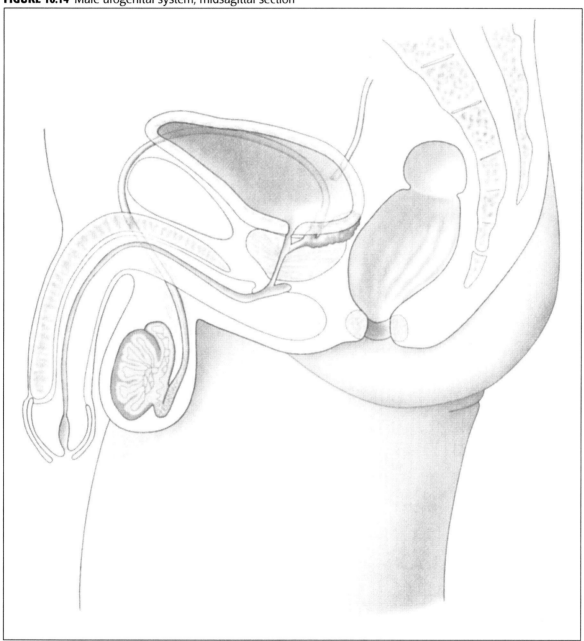

From *Laboratory Guide for Human Anatomy* by William J. Radke, copyright © 2002 John Wiley & Sons, Inc. Reprinted by permission of John Wiley & Sons, Inc.

FIGURE 10.15 Female urogenital system, midsagittal section

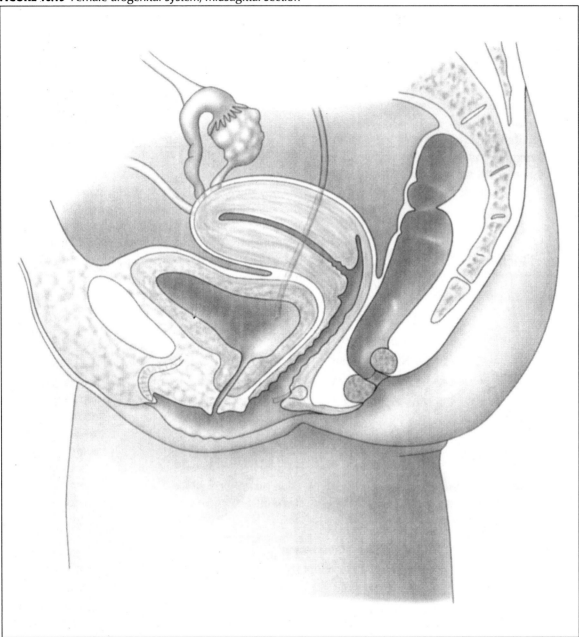